THE GETTYSBURG CAMPAIGN

Major General George Gordon Meade took command of the Army of the Potomac *three days before fighting the greatest battle of the Civil War, Gettysburg. Lee said of his new adversary, "...Meade will commit no blunder in my front, and if I make* ~~haste to take advantag~~

D1457021

GREAT CAMPAIGNS SERIES

GREAT CAMPAIGNS

THE GETTYSBURG CAMPAIGN

June - July 1863

THIRD EDITION

Albert A. Nofi

COMBINED BOOKS
Pennsylvania

PUBLISHER'S NOTE

Combined Books, Inc., is dedicated to publishing books of distinction in history and military history. We are proud of the quality of writing and the quantity of information found in our books. Our books are manufactured with style and durability and are printed on acid-free paper. We like to think of our books as soldiers: not infantry grunts, but well dressed and well equipped avant garde. Our logo reflects our commitment to the modern and yet historic art of bookmaking.

We call ourselves Combined Books because we view the publishing enterprise as a "combined" effort of authors, publishers and readers. And we promise to bridge the gap between us–a gap which is all too seldom closed in contemporary publishing.

We would like to hear from our readers and invite you to write to us at our offices in Pennsylvania with your reactions, queries, comments, even complaints. All of your correspondence will be answered directly by a member of the Editorial Board or by the author.

We encourage all of our readers to purchase our books from their local booksellers, and we hope that you let us know of booksellers in your area that might be interested in carrying our books. If you are unable to find a book in your area, please write to us.

For information, address:
COMBINED BOOKS, INC.
1024Fayette Street
Conshohocken, PA 19428

Library of Congress Cataloging-in-Publication Data
Nofi, Albert A.
The Gettysburg campaign: June-July, 1863/ Albert A. Nofi—revised ed.
p. cm. — (Great Campaigns)
    Originally published: New York, N.Y. : Gallery Books, 1986.
    Includes bibliographical references and index.
    ISBN 0-938289-83-7 :
    1. Gettysburg Campaign, 1863.   I. Title. II. Series.
[E475.51.N63 1993]
973.7'349—dc20   92-34473

First Published, 1986
Revised Edition, 1993
Third Edition, 1997
1 2   3   4   5
Printed in Hong Kong.
Maps by Kevin Wilkins

*For Marilyn J. Spencer
and Lori Fawcett*

In Loving Memory

Acknowledgments

A great many people helped produce this book in one way or another. The listing here is my limited way of offering thanks for their assistance.

Prof. John Boardman and Dennis Casey, with whom I walked the field; Matilda Virgilio Clark, who provided access to certain family records relating to the battle; Morton Berger, historian and curator of the Fourteenth Regiment Armory, Brooklyn, and the members of the 14th Brooklyn Civil War Round Table, who very kindly offered advice and suggestions and access to the historical collection of the 14th Brooklyn; Joseph Sherfy, for information about the family farm at the time of the battle; Daniel Scott Palter and the Staff of *West End Games*, for use of their extensive Civil War collection; Stephan Patejak, Mark Herman, Eric Smith, Daniel David, Bud Livingston, and John Willetts for their books, advice and assistance; Dan Kilbert of New York's Complete Strategist, Richard Berg who helped track down simulation materials; Dr. David G. Martin, for his advice and a copy of his and John W. Bussey's valuable *Regimental Strengths and Losses at Gettysburg* (Hightstown, N.J.: 1986); Robert Capriotti, re-enactor, for his assistance with some of the technical details and some of the pictures; Howard W. Hunter; and the fine people at the New York Public Library, the New York Historical Society, and the Mina Rees Library of the Graduate School of the City University of New York, who patiently tracked down obscure items without knowing why. I am particularly indebted to the officers and men of the recreated *15th New York Engineers, 124th New York Volunteers, Clark's Battery (B, 1st New Jersey Artillery)* and the *1st New Jersey Cavalry* for their valuable demonstrations of kit, equipment, and practice, their cheerful responses to questions, and their devotion to enhancing our understanding of the life of the soldier in the Civil War.

Particular thanks go to Bob Pigeon, Toni Bauer, and John Cannan of Combined Books, who made this volume possible; and to Kevin Wilkins, who did an excellent job on the maps.

Special thanks are due to my wife, Mary S. Nofi, who walked the field as well, and who, along with my daughter Marilyn J. Spencer, suffered through the writing.

A Note on Usage

To simplify matters several ahistorical conventions have been adopted in this work:

1. The identities of Union units are in *italics*.

2. Union Army corps have been designated with Roman numerals.

3. Times have been rendered on a 24-hour basis.

4. Where it is necessary to detail the identity of a Union brigade or division, the form *1st/1st Division/I Corps* has been used rather than the clumsier *1st Brigade* of the *1st Division* of the *I Corps*.

Contents

Maps

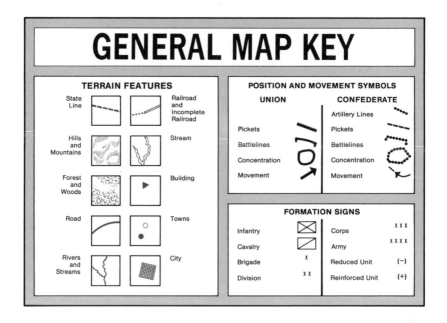

Sidebars

Preface to the Series

*J*onathan Swift termed war "that mad game the world so loves to play." He had a point. Universally condemned, it has nevertheless been almost as universally practiced. For good or ill, war has played a significant role in the shaping of history. Indeed, there is hardly a human institution which has not in some fashion been influenced and molded by war, even as it helped shape and mold war in turn. Yet the study of war has been as remarkably neglected as its practice commonplace. With a few outstanding exceptions, the history of wars and of military operations has until quite recently been largely the province of the inspired patriot or the regimental polemist. Only in our times have serious, detailed and objective accounts come to be considered the norm in the treatment of military history and related matters.

Yet there still remains a gap in the literature, for there are two types of military history. One type is written from a very serious, highly technical, professional perspective and presupposes that the reader is deeply familiar with the background, technology and general situation. The other is perhaps less dry, but merely lightly reviews the events with the intention of informing and entertaining the layman. The qualitative gap between the last two is vast. Moreover, there are professionals in both the military and academia whose credentials are limited to particular moments in the long, sad history of war, and there are laymen who have more than a passing understanding of the field; and then there is the concerned citizen, interested in understanding the phenomena in an age of unusual violence and unprecedented armaments. It is to bridge the gap between the two types of military history, and to reach the professional and the serious amateur and the concerned citizen alike, that this series, GREAT CAMPAIGNS, is designed. Each volume in GREAT CAMPAIGNS is thus not merely an account of a particular military operation, but it is a unique reference to the theory and practice of war in the period in question.

The GREAT CAMPAIGNS series is a distinctive contribution to the study of war and of military history, which will remain of value for many years to come.

Preface to the Third Edition

A number of important changes have been made in this edition of *The Gettysburg Campaign*. Several sections have been completely revised, and there have been numerous smaller changes throughout, in keeping with the most current research on the battle. In addition, the Guide for the Interested Reader has been heavily revised, to keep up with the flood of recent works dealing with the Civil War and the campaign.

Introduction

When the Union began to come apart in late 1860 and early 1861, as the Southern states sought to create a new nation out of the fabric of the old, no one on either side of the secession issue had any firm notion of the consequences. Many, both North and South, believed it would not come to a clash of arms. But the questions at issue were too serious and too emotionally ladened to avoid a struggle. Neither the integrity of the Union nor slavery were issues over which men were any longer willing to compromise. So there was war, the bloodiest and greatest in the American experience.

By the summer of 1863 the American people had suffered through two years of bloody civil war with little to show for it.

For the Union it had been a frustrating time. Great armies had repeatedly marched southwards with hopes high for a swift victory, only to be beaten back in defeat and disgrace. Some, a few, despaired, whether from conviction or political expediency, saying that the war was not worth pursuing any further. But the spirit of the nation remained remarkably firm, sustained by the indomitable will of President Lincoln, and fueled by the Emancipation Proclamation. Despite numerous reverses in the Eastern Theater, there were victories in the West. And the Union remained strong. The enormous agricultural, industrial, and financial resources of the North had been mobilized and production of the munitions of war had soared. Adroit diplomacy had averted foreign recognition of the rebellious South. A powerful navy had been created to deny the use of the rivers and coasts and seas to the South. A great army of some 800,000 men had been raised, trained, and equipped, with some 600,000 seasoned men ready to take to the field. Able commanders were coming to the fore to lead them in increasing numbers. All that was required for victory was time, and will.

The first two years of the war had been frustrating for the Confederacy as well. There had been many great and costly victories, but defeats as well, and neither independence nor foreign recognition had yet been attained. Nevertheless, much had been accomplished. A functioning government had been set up. Whole new industries had been created. A navy had been established and commerce raiders were scouring the seas in search of Federal merchantmen. Most spectacularly, a great army of some 500,000 men had been recruited and trained and equipped and led to victory by a remarkably talented group of generals. It was heady stuff, and Southern morale and determination remained high. To be sure, there were great dangers yet to be faced, but few appear to have doubted the inevitability of victory. However, some discerning individuals had begun to realize that the ability of the South to sustain the war much longer was limited. The nature of war had changed, and changed in ways that were barely perceptible.

What had changed the conduct of war was the Industrial Revolution. It was no longer possible for a war to be decided in one or two great, decisive battles as in the days of Napoleon. The mechanization of industry and agriculture made virtually limitless production of the munitions of war possible, while at the same time releasing enormous numbers of men to fight. Railroads and telegraphs and steamships permitted greater efficiency in the movements of men and material. And the increasing technical sophistication of weapons caused the battlefield to be far bloodier and more dangerous than ever before. Nurtured on the lessons of the Napoleonic Wars, politicians and generals were slow to see the impact of these changes; ordinary citizens had no understanding of them at all. In the new age the society with the greatest resources would be triumphant, given the will to win.

So, appearances aside, the strategic situation of the South in the summer of 1863 was poor. The Confederacy was rapidly approaching the limits of its resources. Little more could be expected from the South's slender industrial base. Manpower was increasingly hard to find. Foreign trade was virtually nonexistent as the Union blockade became ever more effective. Militarily, conditions were deteriorating. In the West, Grant was driving one Confederate army against the fortress of Vicksburg, while Rosecrans prepared to drive another out of Eastern Tennessee and into Georgia. And in the East, despite the reverse at Chancellorsville, the *Army of the Potomac* remained strong and game north of the Rappahannock River. A significant prolongation of the war would not be favorable to the South. Thus was the stage set for the campaign and the battle of Gettysburg.

CHAPTER I

Counsels of War

Concerned with the overall strategic situation of the South, Confederate General Robert E. Lee, commanding the Army of Northern Virginia, began to think about undertaking an offensive into Union territory early in 1863. Such a maneuver had several points in its favor. A continued defensive posture in Virginia would permanently concede the initiative to the Union, for with the spring and summer the *Army of the Potomac* would come south once again and not even Lee and the Army of Northern Virginia could win every battle. Moreover, the present position of Lee's army, just south of the Rappahannock River, was difficult to sustain logistically, particularly as Southern resources were beginning to show the strain of war. Another winter on the Rappahannock line might well prove the ruination of the army. In the North the army would be able to carry off supplies sufficient to enable it to get through the next winter. Invading the enemy's territory would also take the pressure off war-torn Virginia for a while, which might further improve the food situation for the coming year. On a broader strategic level, an offensive would wrest the initiative from the Union. By striking into Pennsylvania, Lee would be in a position to simultaneously threaten both Washington and Baltimore, and, indeed, at least theoretically, even Philadelphia. The *Army of the Potomac* would be unable to avoid a decisive clash and, if defeated, would be too far from Fortress Washington to readily seek shelter, thereby perhaps laying itself open to total destruction. Moreover, Union armies in the West would be forced to

Robert E. Lee on his favorite steed, Traveler. During the winter of 1862-1863, Lee convinced Confederate military leaders to authorize his second invasion of the North. The result was the Gettysburg campaign and one of his worst moments commanding troops.

send troops eastwards, thereby relieving the pressure on the Confederacy along the Mississippi and in Tennessee. Perhaps most importantly, a successful invasion of the North could have far-reaching political consequences. It would strengthen Southern morale, while dealing a severe blow to that of the Union, perhaps encouraging those in the North who were in favor of an immediate end to the war. A successful offensive might also result in diplomatic recognition from Britain and France, and possibly their intervention. In short, an offensive could well win the war.

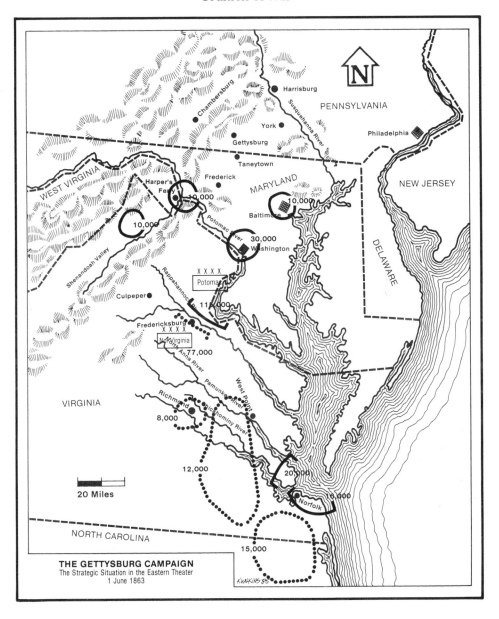

THE GETTYSBURG CAMPAIGN
The Strategic Situation in the Eastern Theater
1 June 1863

Robert E. Lee

Born on a plantation in Virginia, Lee (1807-1870) was the son of Revolutionary War cavalry hero Henry "Light Horse Harry" Lee, whose spendthrift ways resulted in the family living in genteel poverty for many years. Lee graduated from West Point second in the class of 1829. An engineer, he served in garrison, on various civil and military engineering projects, and in Mexico with notable brilliance (one wound and three brevets). He was superintendent of West Point from 1852 to 1855 at a time when two of his sons and his nephew were cadets. He then commanded the *2nd Cavalry* (renamed the *5th* in 1861) in Texas from 1857. In Washington on leave in 1859, Lee commanded the detachment of marines that captured John Brown at Harper's Ferry. Upon the secession of Texas, Lee brought his regiment out of the state despite harassment from secessionist elements. He was then summoned east by Lieutenant General Winfield Scott, the general in chief, and is believed to have been offered command of the Federal armies by Lincoln in April of 1861. He declined the offer and on 20 April resigned his commission, entering Virginia's service as commanding general of state forces.

On 14 May 1861 Lee entered Confederate service as a brigadier general and was promoted full general one month later. He then commanded in West Virginia against George B. McClellan, displaying little evidence of talent as a field commander. He was shortly transferred to command Confederate forces on the Georgia and Carolina coasts and then recalled to serve as military advisor to Jefferson Davis in March of 1862. In this capacity Lee helped plan Jackson's famed Valley campaign and the Peninsula campaign. When Gen. Joseph E. Johnston was wounded defending Richmond during the battle of Seven Pines, Lee was given com-

Other Confederate leaders were thinking about an offensive as well. During the winter of 1862-1863 President Jefferson Davis, a West Point graduate and veteran of the Mexican War (1846-1848), and Secretary of War James A. Seddon had proposed that Lee transfer a portion of his army to the West. This was an idea favored by both General Joe Johnston and General Braxton Bragg, who commanded in Mississippi and Tennessee, General Pierre G.T. Beauregard, usually at odds with the president, and Lieutenant General James Longstreet, commander of Lee's First Corps and one of his ablest subor-

mand of Confederate forces and soon demonstrated remarkable skill as a tactician in the Seven Days Battles (26 June-2 July), beating back McClellan's *Army of the Potomac* from the gates of the city. He then led the Army of Northern Virginia with considerable success during the balance of the Peninsula campaign, and in the Second Bull Run, Antietam, Fredericksburg, and Chancellorsville campaigns, the last of which was his most brilliant tactical success. During the Gettysburg campaign Lee did not perform up to his own standards, failing to properly coordinate the actions of his subordinates and possibly fighting an unnecessary battle, and certainly fighting it in an unnecessary manner.

After Gettysburg Lee continued to lead the Army of Northern Virginia in the long, grinding defensive battles of the campaigns of 1864 and 1865, during which his health deteriorated considerably. In February of 1865 he was made general in chief of all Confederate forces, much too late to influence the course of the war. His surrender at Appomattox on 9 April of that year is generally considered to mark the end of the war.

After the war Lee served as president of Washington College (now Washington and Lee University) until his death. One of the finest soldiers in American history, Lee, who always managed to appear immaculately dressed, regardless of circumstances, looked every inch the fine commander he was. Immensely loved by his troops, who called him "Marse Robert" or simply "Bobbie Lee," Lee was essentially a counterpuncher. A good defensive strategist on a theater level, he seems to have displayed little understanding of the overall strategic situation, though he might have achieved much had he been named general in chief a year or more earlier. Married to a great-granddaughter of Martha Washington by her first marriage, Lee had three sons in Confederate service, two of whom rose to generalships, plus a nephew, and several cousins who did likewise.

dinates. The basic concept was that Lee should accompany a third or a half of his army to the Western Theater, where he would assume overall command, and either conduct operations directly against Grant's armies in the Mississippi Valley, or undertake an offensive from Tennessee into Kentucky and thence across the Ohio River with the intention of cutting the Union in half. However, Lee demurred. He believed that the threat posed by the *Army of the Potomac* was too great to veaken the defense of Virginia by a transfer of part of his nagnificent Army of Northern Virginia to the West. He quite accurately pointed out that rail communications between the two theaters were poor, so that the Union, though operating on exterior lines, would easily be able to counter such a movement. The army was already available for an offensive in the East, where, with distances shorter, and significant objectives closer at hand, it would be more difficult for the Union to react to sudden movements. A vigorous offensive in the East by a reinforced Army of Northern Virginia would be the best way to relieve the pressure in the West. Lee's arguments, and his enormous prestige as the most successful of the Confederacy's generals, overcame all opposition. In the end, after a series of conferences held in Richmond shortly after Lee's spectacular victory at Chancellorsville, President Davis and Secretary Seddon conceded and authorized an offensive into Pennsylvania.

It is unclear when Lee began serious planning for his offensive. As early as February he had ordered maps prepared covering much of eastern Pennsylvania. He worked in secret, as always, though he almost certainly consulted with Lieutenant General Thomas "Stonewall" Jackson and Longstreet. The Chancellorsville campaign briefly interrupted these preparations, but afterwards he immediately resumed planning, while winning support for the offensive. As he envisioned the operation, minor forces would hold the attention of the *Army of the Potomac* on the Rappahannock. Meanwhile, a dense cavalry screen would cover the movement of the Army of Northern Virginia into the Shenandoah Valley. He would then advance up the Valley, his right flank protected by the Blue Ridge and South Mountain chains. Crossing the Potomac in the vicinity of Harper's Ferry, the army would debouch into the Cumberland Valley of Pennsylvania, where it would begin to forage. By keeping his forces well in hand, Lee hoped to be able to isolate and destroy stray elements of the *Army of the Potomac*, thereby weakening the opposition. Should it become necessary to fight a general action, he expected that it would take place in the vicinity of either Chambersburg, York or Gettysburg, modest agricultural and market towns with good road communications. In the event of a successful battle, Lee had a notion to fall on either Washington or Baltimore, possibly forcing a peace. If a general action did not take place by the end of the summer, he would retreat back southwards, having stripped central Pennsylvania bare. To further strengthen his plan, Lee proposed that 20,000 troops be concentrated in the old Rappahannock position under Beauregard, thereby posing an additional threat which might further weaken the *Army of the Potomac.*

Lee's plan was sound, insofar as it went. But it had some flaws. The manpower resources of the South were increasingly slender. It would be difficult enough to bring the Army of Northern Virginia up to strength after its losses at Chancellorsville, let alone find sufficient manpower to reinforce it and to create a new army along the Rappahannock. In addition, the army would be operating in hostile territory far from its base. A defeat could lead to its complete dissolution. Lee would have to exercise very close control over its movements to prevent units from passing beyond mutual supporting distance, while at the same time he would have to disperse his units so that they could forage. The plan left unclear whether Lee would seek to fight a defensive or offensive battle if offered the opportunity for a general action. It also made no provision for the possibility that such an engagement might result in a defeat, leaving the Army of Northern Virginia isolated in the enemy's heartland. Finally, he appears to have assumed that the *Army of the Potomac* would continue to suffer from uninspired leadership. As a result of these flaws, Lee's plan was something of a double-edged blade, able to cut in either direction. By going into Pennsylvania he might well win the war, but he could as easily lose it. Caught up in the preparations for his great offensive, Lee, a bold, pugnacious soldier—and perhaps an overconfident one—saw only the great benefits which it could bring to the South.

The Armies of the Gettysburg Campaign

By mid-1863 the Army of Northern Virginia and the *Army of the Potomac* were old friends. The men of each considered themselves to be the finest troops in the world, and had a soldierly respect for their foes. Each force had its own unique character. The *Army of the Potomac* was always better looking than its rival, which lacked adequate supplies of uniforms, but there was little distinguishing either of them in superior discipline, skill, and courage. Curiously, despite the fact that both armies had been organized and commanded by West Pointers, there were notable differences in the way they were organized. Overall, the Army of Northern Virginia was superior in organization to that of the *Army of the Potomac*, but the latter had a number of organizational advantages of no little consequence.

The basic formation in both armies was the infantry regiment, composed of a headquarters and 10 companies, lettered from "A" through "K," omitting the "J." Occasional regiments had one or two additional companies or one or two fewer, and a small number of Union Regular Army regiments had three battalions of eight companies each. Some regiments were organized into battalions, but most were not. In both armies there were some separate battalions, usually of four companies, though a few Confederate ones had as many as seven.

A Confederate regiment had a colonel, a lieutenant colonel, and a major at headquarters, plus two captains for quartermaster and subsistence duties, a first lieutenant as adjutant, and a second lieutenant as ensign, or color bearer. In addition there was a surgeon major and an assistant surgeon, ranking as a captain. Noncommissioned staff comprised a sergeant major, and three sergeants, one each to deal with quartermaster, commissary, and ordnance matters. Each company had a captain, 3 lieutenants, 5 sergeants, 4 corporals, and from 80 to 112 privates. The organization of a Federal regiment was very similar, with a colonel, lieutenant colonel, and major, plus a quartermaster captain, a lieutenant serving as adjutant, plus a chaplain. There was, in addition, a surgeon ranking as a major or captain, and two assistant surgeons ranking as captains or lieutenants. The noncommissioned staff included a sergeant major, quartermaster sergeant, commissary sergeant, hospital steward, and two musicians, plus the ambulance sergeant and his staff of 12, who were not normally counted on the rolls of the regiment. Each of the companies had a captain, a first and a second lieutenant, a first sergeant and 4 other sergeants, 8 corporals, 2 musicians, and a wagoner, plus from 64 to 80 privates. So at least on paper a Confederate regiment ran slightly larger than a Union one, from 943 to 1,173 officers and men as opposed to from 845 to 1,045 in a Federal outfit. But regiments were rarely full, Northern ones particularly. This was due partially to accumulated losses in battle, but also to political considerations. The Confederacy generally adopted a policy of rebuilding veteran regiments by regular infusions of fresh manpower. The Union generally did not do this, so that veteran outfits gradually ran down to tiny fragments of their former selves. At Gettysburg of those regiments fielding 10 companies, the *61st New York* entered the battle with but 104 men, though the average Union regiment had 300 men, while the weakest Confederate outfit, the 15th Louisiana, had 186, against an average strength of 340.

Above the regimental level both armies were organized into brigades, divisions, and corps, which had very small staffs. In the Army of Northern Virginia these were invariably known by their commander's names, which leads to some confusion particularly when a commanding officer was changed, since the name of his unit would frequently change as well. There was a tendency for Union troops to do this, but generally Union higher formations were more often known by their official designations. However, since all divisions were numbered only within their individual corps, and brigades within their divisions, some confusion is also possible. At Gettysburg Confederate brigades were mostly of four or five regiments, with a few of but three and several with six. Brigades ran between 800 to 2,600 officers and men. Five of the divisions had four brigades, two had five, and one but

three. In addition, each division had a battalion of four batteries of artillery attached. As a result, Confederate division strength was formidable, running between 5,500 and 7,500 men with around 15 to 19 pieces of artillery. Each corps consisted of three divisions, plus two battalions of artillery, for a combat strength of around 21,000 officers and men with about 80 cannons. The Army of Northern Virginia itself comprised three such corps, plus J.E.B. Stuart's oversized cavalry division. Surprisingly, there was no army level pool of artillery, which made it difficult to shift the guns readily, nor were there any combat service troops, so that line infantry units had to serve as engineers at times, and infantry, and cavalry regiments had to provide detachments for military police duties from time to time.

Most Union brigades during the campaign ran four or five regiments, though several had six or more, and a few had three. Brigades ran anywhere from 530 men up to 3,000 due to variations in the number and strength of their regiments. Divisions had mostly two or three brigades, with one having four, which gave them a strength of from 2,400 to 5,100 combatants. Early in the war it had been the practice to add a battery of artillery and some cavalry to each division, but these had been removed long before the Gettysburg campaign. While the loss of the cavalry contingent was not significant, the fact that the division commander had no artillery directly under his control was a serious organizational defect. Two or three divisions plus a brigade of 4 or 5 artillery batteries comprised a corps, which averaged between 9,000 and 13,500 combatants, with from 20 to 46 pieces of artillery with limbers, 200 to 300 wagons and around 50 ambulances, which required 10 to 18 miles of road space if marching in a single column. Altogether, the *Army of the Potomac* had seven infantry corps and one of cavalry. In addition, the army had an *Artillery Reserve*, a brigade of engineers and a mixed brigade under the provost marshal general. The *Artillery Reserve*, of 118 guns and over 2,500 men, gave the army commander a readily deployable pool of artillery, without having to strip batteries from the individual corps. The engineer brigade had troops with technical skills which were just not available among the line troops, even when directed by trained engineer officers. The provost marshal's guard served as field police, thereby relieving individual regiments from having to provide detachments for this purpose. In addition, of course, each Federal brigade, division and corps had its

ambulance corps and medical team, which facilitated treatment of the wounded. These were valuable assets, for which there was nothing comparable in the Army of Northern Virginia. However, despite these valuable advantages, the organization of the *Army of the Potomac* was distinctly clumsy.

Operationally, Meade had nine subordinates under his immediate control, his seven infantry corps commanders, plus that of the *Cavalry Corps* and of the *Artillery Reserve*. The army had a total of 19 infantry divisions and three cavalry divisions. In contrast, Lee had four direct subordinates, one for each corps plus J.E.B. Stuart for the cavalry, and only nine infantry divisions and the cavalry division, thus giving him far fewer people to supervise. So superior was the organization of the Army of Northern Virginia, that in preparation for the campaign of 1864, the *Army of the Potomac* was reorganized, so that it emerged with but four corps, plus the cavalry, artillery reserve and headquarters assets. This reorganization eased the burden of command.

CHAPTER II

The Preliminaries

20 May - 9 June

*T*owards the end of May, Union military leaders began to suspect that Lee was planning something big. Portions of the encampment of the Army of Northern Virginia at Fredericksburg were within view of Union outposts along the Rappahannock. This made it difficult for Lee to make significant preparations for a major undertaking without attracting notice. Lee had also increased his patrols, tightening security to the point where troops were no longer permitted to fraternize across the river. The number of Confederate soldiers deserting to the Union lines had risen—always a sure sign of an imminent offensive—and some of these men provided useful information. Finally, Lee had been pulling in brigades and reinforcements from all over Virginia, a fact which could not go unnoticed. On 27 May Colonel George H. Sharpe, the chief intelligence officer of the *Army of the Potomac*, issued a report in which he indicated that Lee's forces had been significantly increased, detailed the strength and location of Lee's divisions along the Rappahannock, and concluded that the Army of Northern Virginia was preparing for a lengthy campaign away from rail communications. Pointing out that Lee's cavalry, under the redoubtable Major General J.E.B. Stuart, was concentrated at Culpeper Courthouse, considerably to the west of the main encampment, Sharpe decided that Lee planned a movement westwards and

Major General Joseph "Fighting Joe" Hooker. Though decisively beaten by Lee at Chancellorsville from 1-4 May 1863, Hooker remained in command during the early stages of the Gettysburg campaign. His quarrelsome behavior and failure to actively engage the advancing enemy led to his replacement with Major General George Gordon Meade.

then northwards, designed to bring his army against the right flank or right rear of the *Army of the Potomac.* In the middle of restoring his army to fighting trim and attempting to make minor improvements in its organization, Major General Joseph Hooker, still in command despite his humiliating defeat at Chancellorsville, was initially skeptical, but erred on the side of caution and issued preliminary orders alerting units for possible movement. Then, on 28 May, came reports that Confederate skirmishers were reported in the vicinity of Warrenton, some 20 miles north of Culpeper, deep behind his right flank.

Hooker ordered reinforcements to the cavalry screen covering his right, made some minor adjustments to the deployment of his forces along the middle reaches of the Rappahannock and alerted some of his corps commanders to be ready for operations. He made some tentative plans, briefly toying with the idea of falling on Richmond if Lee's movements left it uncovered. Informed of this, Lincoln rightly noted that the principal objective of the *Army of the Potomac* was the Army of Northern Virginia, an observation with which Hooker immediately concurred. He then decided to wait on developments.

Lee hesitated not at all. Although concerned by the presence of some 40,000 Union troops along the Virginia coast under Major General John Dix, he concluded that the latter would pose no serious threat to his rear or to Richmond, should he move northwards. He pressed his preparations, put the finishing touches on his plans, and quietly began to shift units westwards. By calling in virtually every available spare unit in the Eastern Theater, Lee's magnificent Army of Northern Virginia numbered over 75,000 men, fell fighters all. On the morning of Wednesday, 3 June, he made his move. Two divisions of Longstreet's First Corps were the first to move out. The next day a division of one-legged Lieutenant General Richard Ewell's Second Corps marched off, its long column followed the next morning by the other two. In the aftermath of Chancellorsville, and with the knowledge that they were on the offensive, the morale of the troops was very high. Everyone was in good spirits, and the men marched well, with little straggling. The Army of Northern Virginia was on the offensive once more.

Meanwhile, the *Army of the Potomac* had begun to stir. On 3 June Federal balloonists reported that some Confederate camps had been evacuated. Alerted by this and other evidence of activity among the Confederates, Hooker ordered a reconnaissance in force. On 5 June under cover of an intense artillery barrage, some 2,000 men of Major General John Sedgwick's *VI Corps* raided across the Rappahannock at Franklin's Crossing. A brief skirmish ensued and the raiders

Alfred Pleasonton

Born and raised in Washington, Pleasonton (1824-1897) graduated from West Point in 1844, ranking 7th in his class. He served as a dragoon in Mexico (one brevet), on the frontier, against the Seminoles, and in Kansas during the civil disorders preceding statehood. On the outbreak of the war, although only a captain, he led the *2nd Dragoons* (shortly renamed the *2nd Cavalry*) from Utah to Washington. He served in garrison at Washington in the winter of 1861-1862, and as a major in the Peninsula campaign. Promoted brigadier general of volunteers in June of 1862, Pleasonton led a cavalry division in the Antietam, Fredericksburg, and Chancellorsville campaigns with some distinction.

Pleasonton was promoted major general of volunteers and assumed command of the *Cavalry Corps* of the *Army of the Potomac* in June of 1863. He led his corps ably, if not spectacularly, at Brandy Station and through the Gettysburg campaign, and until March of 1864, when he was replaced by Philip Sheridan. He subsequently served in Missouri with some distinction. At the end of the war he was brevetted major general of Regulars. He was offered a lieutenant colonelcy in the infantry, but refused. Unable to secure a cavalry post in which he was not serving under officers junior to him in the volunteer service, Pleasonton resigned from the Army. He was briefly employed in the Internal Revenue Service and later became a railroad president. In 1888 he was granted a pension as a major.

Not a brilliant cavalryman, Pleasonton did his job well, sticking to his instructions, and serving the will of his commanding officer, which was precisely what his Confederate counterpart, J.E.B. Stuart, failed to do during the Gettysburg campaign.

---

fell back, taking with them several prisoners. Information from the raid was ambiguous; Sedgwick reported that the entire Confederate army seemed still to be in its old position. Hooker ordered another reconnaissance in force on 7 June, instructing Major General Alfred Pleasonton, commanding the *Cavalry Corps*, to take his men and some infantry and clear the Confederate cavalry out of the Culpeper area, and ordering elements of Major General George Meade's *V Corps* to support him if necessary. By this time Lee had already concentrated his First and Second Corps in the vicinity of

Culpeper, leaving only Lieutenant General A.P. Hill's reduced Third Corps in the old Fredericksburg position.

By the night of 8 June Pleasonton's 8,000 cavalrymen and 3,000 infantrymen, with six light batteries, were poised along the line of the Rappahannock some eight miles east of Culpeper. Pleasonton formed his troops into two wings. The right wing, under Brigadier General John Buford, comprised the *1st Cavalry Division*, the *Reserve Cavalry Brigade*, and a brigade of infantry, and was positioned at Beverly Ford; while the left wing, under Brigadier General David Gregg, composed of the *2nd* and *3rd Cavalry Divisions* and a brigade of infantry, was at Kelly's Ford, six miles downstream. Buford struck under cover of a morning haze just about dawn on 9 June. The attack achieved complete surprise, although the Confederate pickets, from Brigadier General W. E. "Grumble" Jones' brigade, managed to maintain their cohesion and fell back in good order. Buford's wing drove on towards Brandy Station, a small railroad depot. One of the most confusing engagements of the war ensued. Fierce fighting erupted in front of Brandy Station. J.E.B. Stuart, Confederate cavalry commander, fed his brigades into the fight one after another. Wade Hampton's men came up, followed by those of W.H.F. "Rooney" Lee, second son of the general. These brigades gradually slowed Buford's advance and then, at about 1000, Wade Hampton led them in a counterattack. Buford, who had dismounted almost half his cavalrymen to bolster his infantry, began to fall back under the pressure. Then, at about noon, the bulk of Gregg's wing of the Union *Cavalry Corps* finally came up, having been delayed in crossing the Rappahannock by the late arrival of one division.

Gregg drove on Brandy Station from behind the Confederate right flank. As he advanced, he came under fire from a single light artillery piece on the otherwise undefended Fleetwood Hill, a long, lightly wooded ridge about a half-mile east of the town, upon which was located Stuart's camp. Gregg hesitated, uncertain as to the strength of the forces in front of him. This gave Stuart's assistant adjutant general, Major Henry B. McClellan (a cousin of former Union General

James Ewell Brown Stuart

"Jeb" Stuart (1833-1864) was born in Virginia and graduated from West Point 13th in the class of 1854. Most of his career was spent on the frontier, where he was wounded. From 1855 he was with the *1st Cavalry* (renamed the *4th* in 1861). He was in Kansas during the pre-statehood dispute over slavery and was with Robert E. Lee at the capture of John Brown at Harper's Ferry in 1859. Stuart resigned as a captain on 3 May 1861 and was made a lieutenant colonel by Virginia shortly afterwards. He entered Confederate service soon after and was named colonel of the 1st Virginia Cavalry, with which he served in the Shenandoah Valley and during the Bull Run campaign with great distinc-

tion. He was promoted brigadier general in September and further distinguished himself commanding a brigade of cavalry in various skirmishes through the winter of 1861-1862. During the Peninsula campaign he demonstrated remarkable abilities in reconnaissance, raiding, and screening, and once rode entirely around the *Army of the Potomac.* In July of 1862 Stuart was given command of the Cavalry Division of the Army of Northern Virginia and made a major general. He led his command, later elevated to the status of a corps, thereafter until his death, serving with great skill in the Second Bull Run, Antietam, Fredericksburg, and Chancellorsville campaigns, during the last of

in Chief George B. McClellan), time to call for reinforcements. The 12th Virginia Cavalry Regiment came up, supported by the 35th Virginia Cavalry Battalion, just as Gregg's troopers moved on Fleetwood Hill. As additional forces moved in a thoroughly chaotic melee resulted. Fleetwood Hill changed hands repeatedly in a series of saber swinging charges and countercharges. At one point in the confused action the Union cavalrymen almost overran Stuart's artillery. Pressure on the 12th Virginia was so intense that it broke under fire. The *1st New Jersey Cavalry* made six regiment-sized charges plus numerous smaller ones; other regiments on both sides matched its record. To counter the threat to his rear, Stuart was forced to pull troops away from Buford's front and, as a result, Buford renewed his attack. Stuart gradually fell back, fearing that he might be trapped between two hostile forces.

which he temporarily commanded Second Corps after Jackson and A.P. Hill had both been wounded.

During the Gettysburg campaign Stuart was notably ineffective, taking advantage of some ambiguity in his orders to embark on an unnecessary and unprofitable raid. This raid, undertaken primarily to restore his reputation after the unusual success of the Union cavalry at Brandy Station, effectively deprived Lee of the "eyes" of his army and had much to do with the disastrous failure which resulted. Stuart performed well during the retreat, however, and in numerous skirmishes and engagements along the Rappahannock line in the winter of 1863-1864. He led his corps ably during the battles of the Wilderness and Spotsylvania in the campaign of 1864, but was mortally wounded at Yellow Tavern on 11 May, attempting to block Major General Philip Sheridan's raid on Richmond. He died the next day, greatly mourned.

Stuart was one of the finest cavalrymen in American history, with a remarkable ability to gather information, screen operations, and conduct raids. But he was overly sensitive about his reputation, and perhaps a bit too aggressive, with the result that he had to be issued careful instructions lest he go off on his own, which is precisely what happened in the Gettysburg campaign. Stuart, who wore a beard to hide a receding chin, was the son-in-law of Brigadier General Philip St. George Cooke, a fellow-Virginian who remained loyal to his oath to the Republic, despite the fact that his sons and son-in-law went South.

While the fighting raged in front of Brandy Station, Colonel Alfred Duffie, commanding the Union *3rd Cavalry Division*, was engaged in a hot little fight of his own. Duffie had been split off from Gregg's wing to advance on Stevensburg, some five miles south of Brandy Station, in order to cover the Union left flank. At Stevensburg, he became entangled with about 500 cavalrymen from the 2nd South Carolina and the 4th Virginia. A confused, running fight resulted at the end of which Duffie came away the victor, breaking his opponents, and collecting some 200 prisoners in the bargain. But the little action prevented Duffie from marching on Brandy Station, where his presence might have been decisive.

Fighting around Brandy Station continued into the late afternoon. Stuart had managed to dismount and concentrate most of his forces on Fleetwood Hill, where his slight superiority in numbers gave him a considerable advantage. Plea-

One of the greatest cavalry commanders of all time, James Ewell Brown Stuart. Angered over his near defeat at Brandy Station, Stuart attempted to redeem his reputation by a spectacular, but fruitless, raid around the **Army of the Potomac** *that left Lee without proper reconnaissance.*

sonton had done all that was humanly possible. Many of his men were exhausted, his cavalry and artillery horses worn out, after nearly 14 hours of movement and combat. When Confederate infantry began to arrive, in the form of Major General R.E. Rodes' division of Second Corps, Pleasonton decided to fall back. Stuart's men, equally exhausted, did not pursue as Pleasonton led the *Cavalry Corps* back across the Rappahannock behind a screen of infantry skirmishers.

Brandy Station was a tactical victory for the Confederate cavalry. They had inflicted about 935 casualties on the Union forces, about half of whom were prisoners, while losing some 525, including 200 prisoners, and the Union forces had retired from the field. But morally, Brandy Station was a major victory for the Union cavalry. It had come looking for a fight and done well. For the first time it had stood up to Stuart's

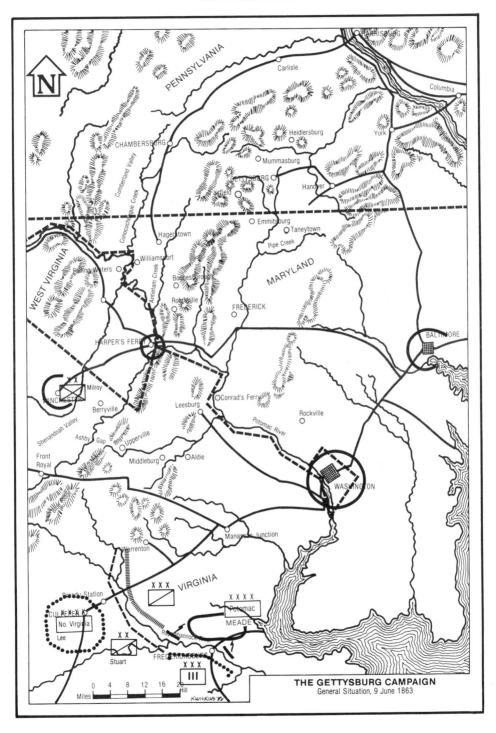

THE GETTYSBURG CAMPAIGN
General Situation, 9 June 1863

vaunted cavaliers on nearly equal terms. Pleasonton's troopers had performed superbly throughout the operation, from their concentration along the Rappahannock on the night of 8 June to their surprise crossing the next morning, to the repeated classic saber charges around Fleetwood Hill and in the hot dismounted fighting that continued through the day, and their withdrawal had been well executed. Stuart and the other Confederate cavalrymen would deny that they had been surprised or that they had been fought to a standstill, but the evidence was clear for all to see. Certainly Lee, who arrived on the field shortly before the end of the battle only to discover that his son had been seriously wounded, could not have been fooled by the improvement in quality of the Union cavalry. Stuart, who gloried in his role as a *beau sabeur*, felt his failure deeply if silently, and the criticism of his subordinates and colleagues rankled. Of course Pleasonton had not actually accomplished his mission, which was to drive the Confederate cavalry out of the Culpeper area. The strength of the Confederate concentration around Culpeper Courthouse and the presence of strong infantry forces in the vicinity prevented this. But his inability to do so provided substantive evidence that Lee was shifting his forces westwards, confirming that a Confederate offensive was imminent.

CHAPTER III

Lee Moves North

10 - 24 June

On 10 June Lee sent Ewell's Second Corps into the Shenandoah Valley, immediately following it up with a brigade of cavalry, and pressing Longstreet's First Corps behind that. Although the weather was oppressively hot and humid, the troops continued to make good progress. Despite several small skirmishes, Ewell covered over 45 miles in two days, reaching the vicinity of Winchester on 13 June, at a time when Hill's Third Corps was still deployed near Fredericksburg, with the result that Lee's army was stretched over 100 miles. But the huge force was well protected. Lee had taken care to move behind a thick screen of cavalry, keeping the Rappahannock River and the Blue Ridge Mountains to his right for cover. Meanwhile, Hooker remained uncertain as to his adversary's intentions; that the Army of Northern Virginia was on the move was certain, but what was its objective? He reacted correctly, if cautiously, shifting his base of operations from Aquia Creek, on the lower Potomac, to the Orange and Alexandria Railroad coming south from Washington, while juggling his army corps westwards. Major General John Reynolds was given charge of Hooker's right wing, comprising his own *I Corps*, plus the *III* and *XI Corps* and the *Cavalry Corps*, with the mission of covering the Rappahannock line. A better general might have guessed at Lee's plans; Hooker, a good corps commander but hardly fit for higher command,

Henry Halleck

A native of New York, Halleck (1815-1872) ran away from home rather than take up farming and was eventually adopted by his maternal grandfather, who saw to it that he received an excellent education (Union College, Phi Beta Kappa). At West Point he was assigned as an assistant professor even before graduating third in the class of 1839. He entered the engineers and helped design the harbor defenses of New York before touring French coast defenses in 1844. He wrote several important works, including *Report on the Means of National Defense* and *Elements of Military Art and Science*, and translated Henri Jomini's *Political and Military Life of Napoleon* from the French. He saw administrative service in California during the Mexican War, emerging with a brevet, and later served as inspector and engineer of lighthouses, and a member of the board proposing fortifications for the Pacific Coast. Meanwhile he found time to help frame the constitution of the State of California and study law. In 1854 he resigned from the Army and founded a major law firm in California. Refusing a judgeship and a U.S. Senate seat, Halleck devoted himself to his profession, writing, business, and the state militia, becoming an authority on min-

waited on events. On 14 June Major General Henry Halleck, general in chief of the Union Army, learned that Ewell was certainly in the Valley, and alerted Hooker. Soon after Hooker was notified that Major General Robert C. Schenck, commanding the Union forces in the Valley, had been informed that Ewell had deployed for battle in front of Winchester on the evening of 13 June. Lincoln then told him that the prevailing opinion in Washington was that Winchester was closely invested by the enemy. Hooker dithered. And as he did, the situation grew worse.

Union forces in the Shenandoah Valley were strong, totaling over 20,000 men. The principal formation was an infantry division of 9,000 men under Major General Robert Milroy. The location of Milroy's division was poor. It was too far south—by over 20 miles—to be readily supported by the 10,000 men stationed at Harper's Ferry and the 1,200 more at Martinsburg. Most of it was at Winchester, a modest town with relatively poor natural defenses, while about 1,800 men were at Berryville, an even smaller town about 10 miles to the

ing and international law and quite wealthy in the process. At the outbreak of the war he was appointed a major general in the Regular Army at the suggestion of Winfield Scott.

Great things were expected of Halleck, but he proved an inept field commander. This fact took some time to become apparent as when he commanded in the West in 1862 his subordinates included Ulysses S. Grant, William S. Rosecrans, and several other able soldiers. Appointed general in chief in July of 1862 his great administrative abilities proved immensely valuable, but he acted much more as a clerk in chief. During the Gettysburg campaign he managed to keep the army supplied and reinforced, but could have done far more to coordinate the activities of the other Union armies. In early 1864 he was replaced by Grant and demoted to the status of chief of staff, a task which he performed to perfection. After the war he held various administrative posts until his death. Halleck's Old Army nickname, "Old Brains," conferred in recognition of his great intellectual achievements, was eventually replaced by "Old Wooden Head." Had there never been a war he might have remained a soldier of great promise, rather than a great disappointment. His brother-in-law, Schuyler Hamilton, grandson of Alexander Hamilton, was also a West Pointer who rose to a major generalcy.

Federal troops in Winchester, Virginia, in 1862. On 13 June 1863 the town was held by 7,000 Federals under the command of Major General Robert Milroy. The Yankees attempted to stand and then retreated from a superior force under Confederate Richard Ewell early the next day. Ewell managed to trap and nearly annihilate the entire command.

Richard S. Ewell

"Dick" Ewell (1817-1872), a native of Georgetown in the District of Columbia, graduated from West Point 13th in the class of 1840, and entered the dragoons. Save for a tour in Mexico (one brevet) his entire service was against Indians in the southwest. He resigned as a captain on 7 May 1861 and promptly entered Confederate service as a colonel in command of a cavalry training camp. He was promoted to brigadier general in June and commanded a brigade at Bull Run. Promoted major general in early 1862, he commanded a division under Thomas "Stonewall" Jackson in the Shenandoah Valley, in the Peninsula, and during the Second Bull Run campaign, where he lost a leg at Groveton, though acquiring a new bride soon after. He returned to duty in May of 1863 as a lieutenant general commanding the late Jackson's Second Corps. His wooden leg prevented him from mounting, so he had to be lifted into the saddle and strapped there during the invasion of Pennsylvania. It was his corps which spread out over much of the central portion of the state in the last week of June. At Gettysburg he ought to have been more aggressive on the first day, and badly mismanaged his attack on Culp's Hill on 2 July by failing to coordinate it with Longstreet's attack further south, possibly costing Lee the battle. After Gettysburg he was again wounded late in 1863, but returned to duty in time to lead his corps in the opening phases of the campaign of 1864. At Spotsylvania a bad fall from his horse rendered him unfit for further field service and he was assigned to command the defenses of Richmond. He was captured during the retreat to Appomattox. After the war Ewell lived in retirement.

Ewell was an inspiring leader, though he hardly looked the part, being bald with a prominent nose, and bulging eyes. After he lost his leg, about which he was not the least bit self-conscious, he got about on crutches, in a travel carriage or, when in action, tied to his horse. Despite his many redeeming qualities, Ewell was unsuited to the command of a corps. Effective when closely controlled, he could not adjust to the freedom which Lee's general instructions gave him. As a result, he was hesitant and relied heavily on the advice of his officers, such as Early's suggestion that he postpone his attack on 2 July.

east. Schenck ignored repeated suggestions by Halleck to correct Milroy's position. Rather than press the issue, Halleck had let it drop. Several small skirmishes occurred between 10 and 13 June as Milroy's outposts fell back before Ewell's advanced guards. As it became clear that Lee was entering the

The commander of Lee's Second Corps, Lieutenant General Richard S. Ewell. Ewell masterminded the destruction of a Federal command in Winchester, Virginia, during the early stages of Lee's offensive. Though a capable officer, Ewell's performance during the battle of Gettysburg would be less than exemplary.

Valley in strength, Schenck had ordered Milroy to fall back. The latter procrastinated, although he did send off his trains. On 13 June Ewell deployed for battle around Winchester with over 14,000 men, while sending a column to block reinforcement from Berryville and another column northwards to seize Martinsburg from the 1,200 men there. For some reason Milroy decided to hold Winchester, although he had but 7,000 men and rations were short.

Ewell struck about an hour before sunset. Twenty pieces of artillery battered the defenders. After 45 minutes, Major General Edward Johnson's division demonstrated against the eastern and southeastern sides of the town while Major General Jubal Early's division attacked on the west and northwest. One of Early's brigades broke into the Union position in a bayonet attack against feeble opposition. Massively supported, the troops pressed on and succeeded in seizing a hill which dominated the defenses. Then night fell, and the fighting ended, both sides having suffered rather lightly. In the darkness, the garrison of Berryville came up, having skirmished with and then eluded the troops sent to cover them. But even with these reinforcements it was clear that the situation at Winchester was hopeless. Milroy decided to retreat along several roads which were reported as still open by his scouts. At 0200 on 14 June his troops moved out in good order, with skirmishers covering the rear. The retreating columns made very good time, quickly covering the four miles to Stephenson's Depot, a hamlet just north of Winchester. There they ran into a brigade which Ewell, with commendable foresight, had placed in ambush. This pinned Milroy in position long enough for additional Confederate forces to come up. Both Milroy and his men performed well, but in the confusion Milroy's rear guard failed to support him. The situation was never in doubt. Even before dawn the command began to disintegrate, with men streaming off in all directions. Milroy managed to keep only 1,500 men together. After an extraordinarily difficult 30-mile march under almost constant harassment from Confederate skirmishers, he brought them safely to Harper's Ferry the next day, even as Lee's advanced guard, Brigadier General A.G. Jenkins' cavalry brigade, was crossing the Potomac nearby. When all the stragglers finally came in, the magnitude of the disaster became clear. Milroy's stand at Winchester had cost nearly half his command and left an enormous store of munitions in Confederate hands. The only benefit of the fiasco was that Hooker had been frightened into action.

Lincoln and Halleck were becoming totally exasperated with Hooker. The latter found it increasingly difficult to get the general to listen to reasonable advice, while Hooker repeatedly complained of "interference" from Halleck. The President resolved the problem by putting Hooker more directly under Halleck's command. Halleck then "advised" Hooker that he should concentrate at Leesburg on the Upper Potomac in a position to strike in any of several directions as the situation dictated. In addition, he was to keep his cavalry as far out as possible in order to discover what the enemy was up to. Hooker therefore ordered his army northwards towards Manassas Junction, moving under the cover of the Bull Run Mountains, while instructing his cavalry to locate Lee's main body.

Numerous small cavalry skirmishes had been occurring with considerable frequency for several days as Union outposts clashed with Confederate outriders. Now Pleasonton ordered his troopers to press harder, seeking information. A

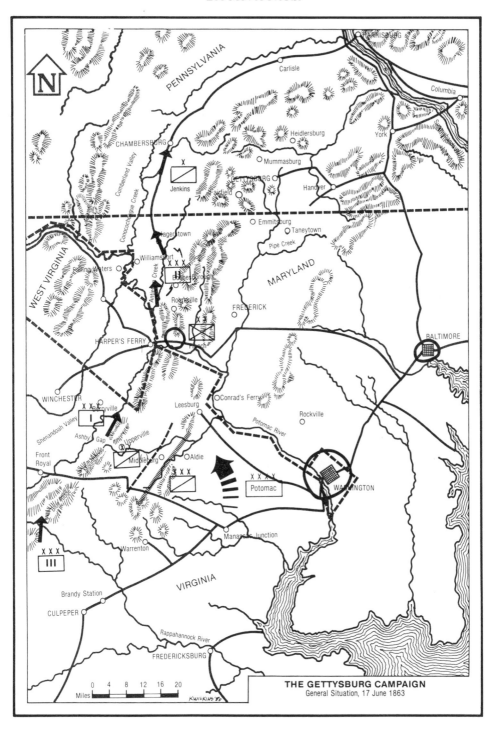

THE GETTYSBURG CAMPAIGN
General Situation, 17 June 1863

Cavalry clash at Upperville, Virginia (21 June 1863), one of a number of skirmishes in which Federal troopers showed themselves to have at last become the equals of their Confederate rivals.

series of larger, hotter clashes resulted, scattered all across the broad valley between the Blue Ridge Mountains, west of which was the Army of Northern Virginia, and the Bull Run chain, east of which was the *Army of the Potomac*. By 17 June J.E.B. Stuart had five brigades covering Lee's right, two along the Rappahannock and three posted at the gaps in the Blue Ridge Mountains. On that morning, Pleasonton dispatched three brigades under Brigadier General David Gregg to reconnoiter towards ·Aldie, a small town from which three of the principal gaps in the Blue Ridge could easily be reached.

Aldie was defended by a single Confederate cavalry brigade, that of Brigadier General Fitzhugh Lee, temporarily under Colonel Thomas T. Munford. A hot little fight ensued, with dismounted cavalrymen trading shots all afternoon. After giving the poorly handled Union troopers a rough time of it for several hours, the defenders pulled out, heading for Middleburg, four miles further west. At Middleburg, Munford ran into about 275 Union cavalrymen under Colonel Alfred Duffie. Duffie's *1st Rhode Island Cavalry Regiment* had just seized the village from a small detachment of Confederate troops when Munford came up. In the face of greatly superior forces, Duffie dismounted most of his command and put up a spirited defense, while dispatching a messenger to

Gregg. Additional Confederate forces were committed. Having held successfully for several hours, Duffie retreated after nightfall. At dawn he discovered himself surrounded. Refusing to surrender, he ordered his men to cut their way out. About 100 succeeded in doing so. That afternoon, in belated response to Duffie's call for assistance, a strong force of Union cavalry drove the Confederates out of Middleburg, and then fell back on Aldie once more. On 21 June, under prodding from Hooker, Pleasonton swept westwards from Aldie with five cavalry brigades and some supporting infantry. Deploying his men in two columns, he quickly pushed the Confederates westwards, fighting a major skirmish at Middleburg and another at Upperville, and drove them on into Ashby's Gap. There Pleasonton stopped when he ought to have pressed on, seeking more information. Still the five days of cavalry fighting between 17 and 21 June proved useful to the Union cause. The Union cavalrymen had again proven capable of standing up to the best the South had to offer. In addition, Pleasonton had inflicted some 770 casualties on the Confederates, including some 250 prisoners, and had taken two guns, at a cost to his own forces of perhaps 880, over 150 of whom had been captured during Duffie's action at Middleburg. Moreover, he definitely ascertained that Lee's entire army was strung-out in the Valley west of the Blue Ridge Mountains and moving northwards. At the same time, and perhaps most importantly, he had totally prevented Stuart from reconnoitering eastwards to the Bull Run Mountains. As a result, Lee was even less well-informed about Hooker's movements than the latter was about Lee's. And it was clear that the Confederate forces were badly sprawled. Lee's advanced guard was well into Pennsylvania, while Hill's Third Corps was still at the southern end of the Valley. Although Hooker had responded sluggishly, he had kept the *Army of the Potomac* well in hand as it moved and it was now concentrated in Northern Virginia, between Manassas Junction and Chantilly, about 20 miles southwest of Washington. At this point a better—or at least bolder—general than Hooker might well have taken advantage of the situation, driving westwards

Skirmish line of Federals armed with carbines. The battle of Gettysburg was preceded by intense cavalry skirmishing.

across the Blue Ridge and cutting Lee in two, as suggested by many, including Lincoln. But Hooker remained cautious. Despite the fact that he was granted considerable authority over all Union forces in the Valley area and Maryland, with detailed instructions, he became involved in petty squabbles over the limits of his authority. Believing himself outnumbered by the enemy, he made repeated demands for reinforcements while denigrating the number and quality of those which he did receive, 7 brigades of infantry and 2 of cavalry, a total of 15,000 field troops, which virtually stripped the Washington garrison of veterans. Despite this increase to his strength, Hooker still felt unready.

Whatever its commander's feeling, the *Army of the Potomac* was prepared to fight. Hooker was a good manager. Supplies were plentiful and morale was good, despite the necessity of marching in a heat wave. Officers were pressing the men hard, but being careful to prevent heat exhaustion and to care for those who succumbed to it. While the *Army of the Potomac* was now closer to Richmond than the Army of Northern

Virginia, Hooker's cautious maneuvering had so far managed to keep the *Army of the Potomac* between Lee and Washington at all times. By advancing in multiple columns along a series of parallel routes—a classically Napoleonic device—he had kept the army fairly well concentrated even while on the move. On 24 June the advanced guard of the army was in the vicinity of Leesburg, with *XI Corps* already crossing the Potomac at Edward's Ferry and *XII Corps* preparing to cross a few miles upstream at Conrad's Ferry. Covered by Pleasonton's cavalry on its left flank, the balance of the army was moving up rapidly, with the rear guard less than 20 miles to the south.

The Army of Northern Virginia was ready too. By 24 June, save for three cavalry brigades, Lee had brought virtually all of his elated troops across the Potomac. Ewell's Corps was well inside Pennsylvania, already foraging heavily in the vicinity of Chambersburg, while Longstreet was resting at Hagerstown in Maryland, and Hill's Corps was on the march across the old Antietam battlefield. Nevertheless, Lee was uneasy. The week of cavalry skirmishing had brought in little information and he had no idea of the location of Hooker and the *Army of the Potomac*. On 22 June he had instructed Stuart to leave two brigades guarding the passes into the Valley and to shift the balance of his cavalry to the head of the army as quickly as possible and by the most direct route. Unfortunately, as written the orders were vague, a not uncommon characteristic of Lee's staff work. Stuart chose to interpret them as permitting him to ride completely around the Union army, rather than as a directive to interpose himself between the two armies. He had performed this feat in earlier campaigns, to his greater glory but with little profit to the Army of Northern Virginia. As a result, on 24 June Stuart vanished, taking with him three brigades of cavalry and a battery of horse artillery. Meanwhile, Lee developed his plans further. Over the next few days Lee intended to spread his troops widely over the rich farms and industrious villages of central Pennsylvania, reaping a rich harvest.

Pennsylvania was virtually bare of defenders to meet the enemy invasion. Concern had been rising for some days. A home guard had been organized with the help of the War Department, and Governor Andrew G. Curtin had called upon the men of the state to enlist. Major General Darius N. Couch, who had ably led a corps in the *Army of the Potomac* for over a year until Hooker's performance at Chancellorsville led him to request a transfer, was appointed commander of the *Department of the Susquehanna*, covering the eastern half of the state. Recruiting for this home guard proved difficult, for Pennsylvania's finest sons were already in the field. Moreover, time was short, there were many technical problems, and equipment was limited. There was considerable war weariness as well, and not a little hostility towards the war effort, for there were many pacifistic Quakers and Mennonites in the state. As a result, on 15 June Lincoln issued a proclamation calling for 100,000 volunteer militia from Pennsylvania, Maryland, West Virginia, and Ohio to serve during the emergency. On 16 June all Northern governors were requested to provide special limited-service volunteers as well. Only New York and Rhode Island responded, the former with an immediate offer of 10,000 three-months militia, mostly from New York City. Couch was ordered to enroll and equip these men as soon as they reached Pennsylvania. Meanwhile he arranged for the fortification of the eastern side of the Susquehanna and for the bridgeheads at Harrisburg. The troops began to move. By 24 June perhaps 15,000 militiamen were concentrated in the central portion of the state, notably about Harrisburg, though many more enrolled as the scale of the Confederate incursion became increasingly obvious. They were not the best troops in the world, but they would fight if they had too, and Couch would do what he could with them. Despite these defensive measures though, panic began to spread as Lee's advanced guards moved across the state.

CHAPTER IV

The Armies Close

24 - 30 June

Lee concentrated his army in the vicinity of Chambersburg, where, impressed by his reputation and noble appearance, a local matron ceased defiantly waving the "Stars and Stripes" long enough to cry "O, I wish he was ours!" From Chambersburg Lee moved detachments all over Pennsylvania between North Mountain and the Susquehanna River. Many of the citizens had already fled, carrying their goods with them. Those who could not, or would not, were forced to deal with the Confederate quartermasters. The spirit of the people was sullen. There was little overt resistance, but little willing cooperation. The primary motivating force was fear and it was particularly evident among the black citizens of Pennsylvania and Maryland. Over 10,000 blacks lived in the areas through which the Army of Northern Virginia was moving. Many were long-term residents, freemen from birth. Others were recent fugitives from slavery. Most joined the flood of refugees fleeing before the advancing Confederate forces. Their fear was justified. Several Confederate commanders, including Longstreet, had given orders that their troops were to apprehend fugitive slaves and return them to their owners. Lee, who was certainly aware of these orders, made no attempt to counteract them. As a result, hundreds of unfortunate blacks were seized in places like Hagerstown, Mercersburg, McConnellsburg, and Chambersburg, and sent South,

Having stripped off their shoes and trousers, Confederate infantrymen prepare to ford a stream during Lee's great offensive.

despite efforts by local white citizens to aid them. Meanwhile the Army of Northern Virginia proceeded to plunder Pennsylvania, favored by unseasonably mild temperatures.

Forage was good and Lee issued careful instructions as to its collection in order to gain the maximum benefit for the Army of Northern Virginia. Private citizens were not to be molested. Specific officers were assigned to levy requisitions on each town, hamlet, and farm, detailing specific quantities of particular products which were to be taken under penalty of retribution. An enormous amount of material was seized, including foodstuffs, cattle, medicine, clothing, horse furniture, and other supplies. No horses were permitted to slip from the quartermaster's grasp, and literally thousands were seized. All goods taken were paid for in Confederate currency. Those refusing to cooperate or concealing goods found their goods seized, in return for a claim receipt against the Confederate government. In practical terms there was little difference, as both the money and the receipts were worthless paper in Pennsylvania, and not much more valuable in the Confederacy. Officially, care was to be taken to see that no persons were actually left to starve, but the requisitions amounted to confiscation and hardship for some was inevitable, for most towns were visited more than once and the seizure of horses hampered gathering the harvest. Confederate behavior was largely polite, if forceful. The troops were generally kept well in hand though there was, to be sure, some casual looting, and some citizens were mistreated by individual Confederate soldiers. Moreover, some Confederate officers ignored orders and sought to punish the Yankees. In this fashion, in complete disregard for Lee's instructions, Jubal Early improperly put to the torch an iron foundry belonging to Representative Thaddeus Stevens, one of the most vehement abolitionists in the House, who happened to make his home in the small town of Gettysburg. As the looting of Pennsylvania proceeded, the armies began to converge.

By 28 June the Army of Northern Virginia was dispersed in an arc stretching some 72 miles from Chambersburg, with sizable contingents at Carlisle and York and cavalry posted along the line of the Susquehanna and westwards from Chambersburg. A number of small skirmishes with local militia had taken place, but nothing to indicate the presence of the bulk of the Union forces. Few of the militia units put up much resistance, wisely choosing to withdraw rather than confront Lee's seasoned veterans. Meanwhile, Lee realized that his army was dangerously overstretched. He was still completely out of touch with Stuart's cavalry and was very much in the dark about the movements of the *Army of the Potomac*.

Hooker had taken the *Army of the Potomac* across the river whose name it bore by 28 June, and had concentrated it in central Maryland, in an area stretching about 20 miles westwards from Frederick. He had conducted his movements

51

carefully, keeping his cavalry well out in front, to prevent the enemy from receiving any intelligence while seeking maximum information himself. His new position was excellent, for he could cover all of Lee's optional movements simultaneously, whether the latter chose to retreat up the Shenandoah Valley back to Virginia, or to fall on Washington or Baltimore. However, Hooker's movements were timid. An enemy army was plundering the heart of Pennsylvania and his army was posing no direct threat to it. A fine administrator, a good strategist, and an able corps commander, Hooker was unsuited for independent command and was beginning to lose his nerve again. His corps commanders had lost confidence in him. In the last few days of the campaign a galaxy of generals had passed through Washington, honestly if insubordinately expressing their reservations about Hooker. Meanwhile, Hooker alienated Halleck, with whom he had never been on friendly terms, by refusing to inform him of his plans. He exacerbated relations with the general in chief further by demanding yet more reinforcements, although already he was in command of over 100,000 field troops. When informed that there were no more troops to be had, he urged the evacuation of Major General William Henry French and the 11,000 men garrisoning Maryland Heights, a strong, well-fortified position which dominated Harper's Ferry and stood athwart part of Lee's line of retreat. Halleck once more turned him down. On 27 June Hooker offered his resignation, believing that this would force Lincoln to give him a freer hand. But the President had by now lost confidence in the general as well. Aware that a great battle was imminent, Lincoln realized that Hooker was not the man for the job. The resignation was accepted. Early on the morning of 28 June, Major General George G. Meade, commanding *V Corps*, was roused from his bed by one of Halleck's staff officers, who handed him a presidential order giving him command of the *Army of the Potomac*. Meade's reaction was, "Well, I've been tried and condemned without a hearing, and I suppose I shall have to go to execution." Then he set about assuming command. Time was short, for it was clear a great battle was in the offing. That very day *The New York Times* had observed that "the return game between the *Army of the Potomac* and the Army of [Northern] Virginia may be played this week."

Meade was a good officer, with a solid, if unspectacular reputation. He had done well with a division at Antietam and Fredericksburg, and had led his corps ably at Chancellorsville. The other corps commanders in the army liked and respected him, and appear to have formed a cabal with the intention of securing the command for him. The rank-and-file, however, found him unimpressive. Nevertheless, although neither as flamboyant nor as brilliant as Hooker, he was a careful, determined, conscientious, loyal, and courageous commander, of whom Lee would remark, "...Meade will commit no blunder in my front, and if I make one, he will make haste to take advantage of it." He had other qualities which made him attractive to Lincoln: born in Spain, he was considered technically disqualified from running for president, and he made his home in Pennsylvania, and as the President put it, "Meade will fight well on his own dung hill." Meade's instructions were clear: all troops in the area of operations were to come under his command and, save that he had to maintain the army in a position to cover both Washington and Baltimore from the enemy, he was given a free hand, even to the extent of replacing any officer in the army. Meade took charge immediately and with notable efficiency. He spent the first few hours reviewing the strategic situation and conferring with Hooker. Even as he formally took command at about 0700 hours, retaining most of Hooker's staff, he was making plans to get the *Army of the Potomac* on the road once more. He had worked fast, and telegraphed a tentative plan to Washington along with his acceptance of command. His plan was simple. He would advance northwards towards the Susquehanna, keeping Lee away from Washington and Baltimore, to concentrate along Pipe Creek, a highly defensible stream about five miles south of the Mason-Dixon Line, and then seek a general action at the earliest opportunity.

George G. Meade

Meade (1815-1872) was born in Cadiz, Spain, where his family, American citizens with roots in Pennsylvania, had mercantile interests. He graduated from West Point 19th in the class of 1835 but resigned from the Army the following year to practice civil engineering. In 1842 he reentered the Army as a lieutenant of Topographical Engineers and was thereafter employed almost continuously in a variety of navigational engineering projects, including lighthouse construction, coastal surveys, and port development, save for a tour of duty in Mexico, where he won a brevet. In 1861 he was named a brigadier general of volunteers at the urging of Governor Andrew G. Curtin of Pennsylvania and given a brigade of Pennsylvania troops. He led his brigade in various corps with distinction in the Peninsula campaign, where he was twice wounded on the same occasion, and at Second Bull Run. He was then given a division under Hooker in *I Corps*, leading it at Antietam and Fredericksburg. At the end of 1862 he was made a major general and given *V Corps*, which he led at Chancellorsville. Hooker's manifest inability to lead the *Army of the Potomac* in the early days of the Gettysburg campaign caused Meade to be appointed to its command on 28 June after Major Generals John Reynolds, John Sedgwick, and Henry W. Slocum, all senior to Meade, indicated a wilingness to serve under him.

Orders began to flow. Outlying contingents of the army were ordered to concentrate on Frederick. Major General George Sykes was entrusted with Meade's old *V Corps*. After consulting with Pleasonton, Meade reorganized the *Cavalry Corps*, with additional manpower and more horse artillery, and captains George A. Custer, Elon Farnsworth, and Wesley Merritt jumped to one-star rank in the volunteer army and given cavalry brigades. Some of the cavalry was sent off to chase Stuart, then about 10 miles northwest of Washington. After long and careful thought Meade ordered Major General French, commanding at Maryland Heights above Harper's Ferry, to pull his men back to Frederick, in order to guard the rear and flank of the army as it moved north. Meade kept Halleck informed of all his actions, and the latter concurred in every instance, limiting his comments to general observations and an expression of concern that military stores not be

Meade immediately had to confront Lee's Army of Northern Virginia in Pennsylvania, a task which he did with considerable ability given the circumstances—the battle of Gettysburg began just two days after Meade assumed command—inflicting a serious defeat on Robert E. Lee. His movements thereafter have been criticized, but the balance of opinion seems to agree that his caution was well founded. Meade remained in command of the *Army of the Potomac* through the end of the war, serving directly under Lieutenant General Ulysses S. Grant in the campaigns of 1864 and 1865.

After the war Meade was named a Regular major general and served in a variety of administrative posts until his death. By no means a brilliant soldier, Meade nevertheless understood his mission, was careful of his troops, was open to advice, and was willing to fight to the utmost when necessary, precisely the qualities which earned him the command of the *Army of the Potomac*, in which he served ably if not spectacularly. A modest, unassuming man, he could display considerable anger if confronted by stupidity or ineptitude. Two of Meade's sisters married Southerners and were ardent secessionists, their husbands serving in the Confederate forces. The husbands of his other three sisters all served the Republic and his brother was a captain in the U.S. Navy. Meade's son served on his staff at Gettysburg. Mrs. Meade's sister was married to Henry A. Wise, who as governor of Virginia had signed John Brown's death warrant and was a Confederate brigadier general during the war.

---

left for the Rebels under any circumstances. Benefiting from some remarkably accurate intelligence provided principally by loyal citizens in the Rebel-occupied regions of Pennsylvania and Maryland, Meade was able to prepare meticulously detailed plans for the movement of each corps. The movement began on 29 June and was almost a textbook study.

The army moved well screened by its cavalry. Buford's *1st Cavalry Division* covered the left and front, while Brigadier General Hugh Judson Kilpatrick's *3rd Cavalry Division* covered the right flank and Gregg's *2nd Cavalry Division* covered the right rear and chased Stuart, who by now was some 25 miles west of Baltimore. The advanced guard comprised *I Corps* and *XI Corps* under the general direction of the very able Major General John Reynolds of *I Corps*. Behind these moved the main body of the army, with *II Corps*, *III Corps*, *V Corps*, *XII Corps*, the *Artillery Reserve*, and with *VI Corps* in the

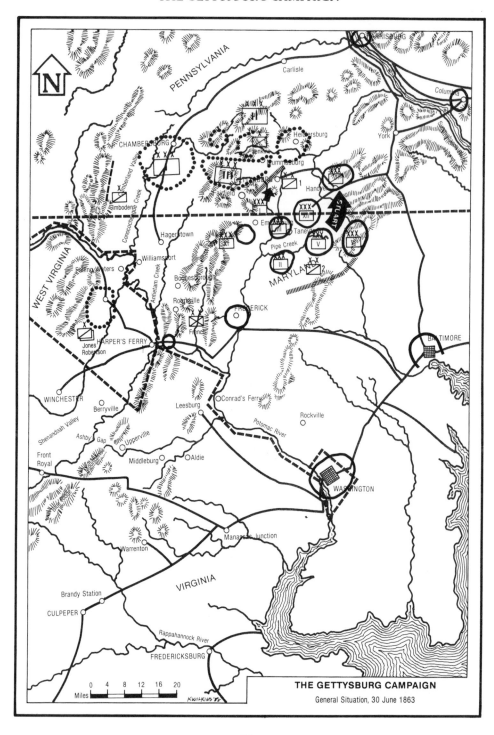

THE GETTYSBURG CAMPAIGN

General Situation, 30 June 1863

rear acting as a general reserve. Despite considerable straggling, and some crossing of columns, it was probably the most well organized and most carefully conducted movement that the *Army of the Potomac* had ever undertaken. Though the average advance on 29 June was about 20 miles, some units exceeded this. Buford's cavalry advanced over 35 miles, a feat which could be matched by some of the infantry, notably Major General Winfield Scott Hancock's *II Corps*, which covered 32 miles, mostly at night and with minimal straggling. Morale was good, and the troops were very determined, particularly the Pennsylvanians, who constituted nearly 40 percent of the army (68 infantry and 9 cavalry regiments, plus 5 batteries of artillery). On 30 June, Meade began to push the right of the army forward somewhat, so that by nightfall his front ran roughly southwest to northeast a few miles north of Pipe Creek, while the rear was just south of the creek. It was an excellent position in which to meet any possible Confederate advance. The bulk of the army was concentrated within a dozen miles of Taneytown, just below the Pennsylvania-Maryland border, so that no corps was more than a few hours march from any other. His cavalry was far ahead. Kilpatrick was at Hanover, where he had a hot skirmish with Stuart's cavalry, and Buford's pickets were about four miles beyond Gettysburg, a small market town. In this position Meade rested his army. He instructed his corps commanders to issue extra ammunition and rations, and to prepare their men for battle. He was careful to inform them of his plans. For 1 July the general prepared two sets of plans: one proposed a withdrawal to the Pipe Creek position should the enemy fall on the *Army of the Potomac*, while the other assumed a less aggressive foe, and envisioned a cautious advance in the direction of Gettysburg. If it came to a fight over the next few days, Meade was ready and so was the *Army of the Potomac*.

The Army of Northern Virginia was also on the move during the last days of June. Lee had bungled badly. By allowing Stuart a free hand, he had deprived himself of one of the finest reconnaissance forces in the world. Moreover, he had also failed to use the cavalry brigades which Stuart had left behind. The result was that as late as 28 June he remained ignorant of the movements of the *Army of the Potomac*, yet he continued to leave his army spread out over some 2,000 square miles of Pennsylvania. Then, on the night of 28 June, he received some disturbing information. A spy dispatched to Washington by Longstreet on 24 June, had returned to the army with the news that the *Army of the Potomac* was in Maryland in force. In haste, and somewhat carelessly, Lee began to issue orders to pull in the more dispersed elements of the army. Ewell, commanding the most exposed part of the Army of Northern Virginia, the Second Corps, was disappointed when orders to pull back were received. He had just fought a modest skirmish before the defenses of Harrisburg and believed that he could take the Pennsylvania capital at his leisure, bringing much glory and loot to Confederate arms. Nevertheless, he obeyed orders, or at least tried to, for he received two different sets of instructions. The first set, issued on the night of 28 June, directed him to march on Chambersburg, while the other, issued the following morning, ordered him to concentrate at Heidlersburg. Other commanders had similar experiences. As a result, when the Army of Northern Virginia began to move on 29 June, there was considerable confusion. Units already on the road were forced to reverse their direction. Some units found that roads assigned to them had been assigned to other units as well. Johnson's Division of Second Corps ran into part of Longstreet's First Corps, resulting in a massive traffic jam which was resolved with difficulty. Most units did not have such serious problems, but there was considerable confusion. Fortunately morale was very high and the troops performed splendidly, despite their grumblings. When the movements were completed on 30 June, the army was concentrated in an area 30 miles long by 10 wide, between Heidlersburg and Chambersburg. The army was still somewhat dispersed, and there was a good deal of intermingling of units, but it was the best that could be done in the circumstances, and Lee planned additional movements over the next day or so to concentrate the army in the vicinity of Cashtown and Gettysburg. Scouts were sent out and began to bring back useful information, including the news that Meade had assumed command of the *Army of the Potomac*, but nothing substantial, such as the location of the enemy. Nevertheless, Lee appears to have felt confident that he could deal with any eventuality. Meanwhile, some of the troops were still on the march. The evening of Tuesday 30 June was warm but not uncomfortably so, and the troops of Brigadier General James J. Pettigrew's brigade of Major General Henry Heth's division were still on the road, advancing on Gettysburg behind a thin line of skirmishers. At Gettysburg Pettigrew, whose men needed shoes, intended to levy yet another contribution on the already well-picked over town and its 2,500 inhabitants. In the gathering darkness, Pettigrew's skirmishers ran into some Federal cavalrymen along the Chambersburg Pike about four miles northwest of Gettysburg. They were from Buford's *1st Cavalry Brigade*, placed on outpost duty earlier that evening. Pettigrew halted his men about a mile from the Union vedettes. Both Buford and Pettigrew dispatched messengers to the rear. The two armies had made contact. The great battle which everyone had been anticipating and dreading for days was imminent.

A view of Gettysburg looking southeastwards from Oak Hill. The skirmishing which preceded the battle took place to the right and rear of this position.

CHAPTER V

Meeting Engagement

1 July

*T*he first shots of the battle of Gettysburg were fired not long after daylight on the morning of 1 July by Union cavalrymen on picket duty along the Chambersburg Pike just west of a small stream named Willoughby Run. Almost immediately, Henry Heth started his division on the road towards Gettysburg, and at about 0500 elements of Pettigrew's Brigade formed a 2,500-yard-wide skirmish line and began to advance slowly. Beyond the Rebel skirmishers, the Union cavalrymen could discern considerable activity as the balance of Heth's Division came up. The pickets formed a slender skirmish line under cover, with no more than 1 man for every 10 yards of front. At first it was a desultory action. But as he became aware of how thin the Federal forces were to his front, Heth pressed them harder. The Union troopers began to fall back slowly. At about 0800 their brigade commander, Colonel William Gamble, was informed that his outpost line was under considerable pressure. Conferring with Buford, the latter immediately ordered him to put 1,200 of his men and a battery of artillery into battle line along Herr Ridge, a modest rise about 2 miles west of Gettysburg, which offered some natural advantages. As a precautionary measure, Buford's other brigade, under Colonel Thomas E. Devin, was posted to cover the approaches further to the north. Gamble's troopers went into action rapidly. Though greatly outnumbered, they

John Buford

A native of Kentucky, Buford (1826-1863) was raised in Illinois. In 1848 he graduated from West Point 16th in his class and thereafter saw much frontier duty in the *2nd Dragoons*, renamed the *2nd Cavalry* in 1861. In the early part of the war Buford served as a major and inspector general of the defenses of Washington. Major General John Pope appointed him a brigadier general in July of 1862 and gave him the *Reserve Cavalry Brigade* of his *Army of Virginia*. Buford served with great distinction during the Second Bull Run campaign and was so badly wounded as to be reported as dead. Serving as chief of cavalry for the *Army of the Potomac* during the Antietam and Fredericksburg campaigns, Buford was returned to his brigade when Hooker reorganized the army in early 1863. He served ably in the Chancellorsville campaign, but it is for his services on the first day at Gettysburg that he is chiefly distinguished. In the autumn of 1863 Buford took sick with typhoid fever and died soon after, shortly after being appointed a major general of volunteers. One of the finest cavalrymen in American history, an admiring observer noted that "Buford always knew what to do when the going got rough." His brother, Napoleon Bonaparte Buford, also rose to a major generalship, serving principally in the West. He had a cousin, Brigadier General Abraham Buford, in Confederate service.

Brigadier General John Buford led the cavalry forces that fought a delaying action at Gettysburg against Henry Heth's Confederate division until Union infantry forces arrived. A superb cavalryman, sadly Buford would take ill and die on 16 December 1863.

Yankee troopers of Brigadier General John Buford's 1st Cavalry Division skirmishing before Gettysburg on 1 July 1863. Note the men holding the horses: in dismounted action one man in four had to be detailed to this duty.

had a strong position. Moreover they carried Sharps carbines, with an effective rate-of-fire three times that of an ordinary rifled musket. Faced with more substantial opposition, Heth proceeded cautiously, putting his men into six columns, three of which advanced on either side of the Chambersburg Pike. The Union cavalrymen maintained their line for nearly an hour before being forced to fall back about a mile to the partially wooded McPherson's Ridge, an even better position. Here the men held tenaciously awaiting reinforcement.

Major General John Reynolds, commanding both *I Corps* and the advanced guard, was the first reinforcement Buford received; he arrived with his staff a bit after 1000. Conferring with Buford, Reynolds commended him on the excellence of his dispositions and his choice of ground, which was naturally quite strong and dominated any possible Confederate advance on Gettysburg from the west. It looked to Reynolds like the ideal place to begin a major action with the enemy, with plenty of room to fall back later to even better positions. Reynolds ordered Buford to hold at all costs. He dispatched a message to Meade, informing him of the situation and concluding, "I will fight them inch by inch, and if driven into the town I will barricade the streets and hold them as long as possible." Then, after sending messages to Howard of *XI Corps*, Slocum of *XII Corps*, and Sickles of *III Corps* requesting

Henry Heth

A native Virginian, Heth (1825-1899) was refused an appointment to Annapolis and as a result went to West Point, graduating in 1847 at the bottom of his class. He served in Mexico, on the frontier and in garrison, rising to captain before resigning in April of 1861 to enter Confederate service in the same rank. He was shortly after commissioned colonel of the 45th Virginia and served in West Virginia in late 1861. Early the following year he was promoted brigadier general and held an administrative post in western Virginia. He then went West, rising to division command during Braxton Bragg's Kentucky campaign. Early in 1863 he was transferred to the Army of Northern Virginia at Lee's request and was given a brigade in A.P. Hill's division, leading it during the Chancellorsville campaign, where he took over a division. On the morning of 1 July skirmishers from his division opened the ball at Gettysburg. Heth handled his forces badly, losing heavily, and being himself severely wounded, though able to return to duty within a few days. After Gettysburg he participated with his division in all subsequent campaigns of the Army of Northern Virginia, from the Wilderness to Appomattox.

After the war he was in the insurance business, a surveyor, and served in the Office of Indian Affairs. Heth, said to have been the only person in the Army of Northern Virginia whom Lee addressed by his first name, was an able commander, but had to be closely supervised. During the Gettysburg campaign he twice blundered into engagements, at Gettysburg itself on 1 July and at Falling Waters on 14 July, when he was supposed to be covering the Confederate retreat.

---

their immediate assistance, he rode back to hurry the 3,700 veterans of his *1st Division*, under Brigadier General James S. Wadsworth, about 30 minutes march away. Meanwhile action to Buford's front grew heated, as Heth attacked on a two brigade front. Buford's men began to fall back. Then, shortly after 1030, Reynolds brought up his leading elements. Barely in time, Reynolds got the *2nd Maine Battery* into position, and then the *56th Pennsylvania* of Wadsworth's *2nd Brigade*, which was followed by two other regiments north of the Chambersburg Pike, while two more deployed to its south. Buford's tired troopers took up positions on the flanks. By 1045 the Federal line had been strengthened and lengthened some-

General Confusion

The *Army of the Potomac* had 65 generals, the Army of Northern Virginia 53. There were fourteen major generals of volunteers in the *Army of the Potomac*: the army commander, the chief of staff, eight corps commanders, and four division commanders. There were also 51 brigadier generals. They commanded 18 divisions, including the three of cavalry. Of 50 infantry and six cavalry brigades, half of each were led by brigadier generals and half by colonels. There were also five brigadiers at army headquarters in various staff positions. In contrast, the Army of Northern Virginia had one full general commanding, plus one lieutenant general for each of the three corps. There were eleven major generals, one for each division, including the cavalry, plus a spare at headquarters. Of 37 infantry brigades, brigadiers commanded all but six, which were led by colonels, while in the cavalry there were six brigadier generals leading brigades and one colonel. There were about 1,430 troops for each general in the Army of the Potomac, and some 1,300 for each general in the Army of Northern Virginia. Adding in the colonels serving in command of brigades, the ratios fall to about 1,010 men for each of the 92 general officer posts in the Army of the Potomac

and 1,165 for each of the 60 general officer posts in the Army of Northern Virginia. These figures may seem excessive by modern standards, but were quite normal for the times. Communications were much more difficult and it was necessary to have more rank at lower levels. Indeed, the somewhat lower proportion of troops to generals probably gave the *Army of the Potomac a useful,* if small, advantage, since it was possible to supervise the troops more closely. In 1863 generals actually led their troops into action, as demonstrated by their casualty rates during the battle: five Confederate and four Union generals were killed or mortally wounded during the battle, and a dozen Confederate and thirteen Union generals were less seriously wounded, making for a Confederate general officer casualty rate of 32 percent and a Union one of 26 percent. Of colonels leading brigades, the Union lost four killed and seven wounded, the Confederacy six wounded. So 30 percent of the Union officers holding brigade or higher conmmands became casualties, as did 38 percent of their Confederate counterparts, exclusive of men taken prisoner. Being a general in the Civil War was not a low risk proposition.

John F. Reynolds

A native of Lancaster, Pennsylvania, Reynolds (1820-1863) attended various schools before entering West Point, from which he graduated 26th in his class in 1841. He served against the Seminoles, on the frontier, in Mexico (two brevets), on exploration and survey duty in the West, and was in the "Mormon War" before being named commandant of cadets at West Point in 1860. In the spring of 1861 he was named lieutenant colonel of the newly activated Regular *14th Infantry Regiment* and that August was promoted to brigadier general of volunteers. He commanded a brigade in the defenses of Washington and in the *Army of the Potomac* during the Peninsula campaign, in which he was captured. Exchanged in August of 1862, he commanded a division of the famed *Pennsylvania Reserves* in *III Corps* during the Second Bull Run campaign before being assigned to command the militia called out in Pennsylvania during the Antietam campaign. Reynolds was promoted major general of volunteers and given *I Corps*, which he led at Fredericksburg and Chancellorsville. He is believed to have refused the command of the *Army of the Potomac* in favor of George G. Meade, who was his junior.

Meade made Reynolds commander of the right wing of the army and at Gettysburg his prompt and effective action on the morning of 1 July was important in determining that the ensuing action would be the decisive clash of the Pennsylvania campaign. He fell as he rode the lines, shot by a Rebel marksman, not 50 miles from his home. Reynolds was not only a fine general with an instinct for doing what was right in the midst of battle, but he looked like one, standing six-feet with dark hair and eyes, and a frame which wore a uniform well. His brother William Reynolds had a distinguished career in the navy, rising to rear admiral.

what with about 2,000 infantrymen and six 3-inch field guns stretching from the McPherson farm northwards for about 2,000 yards across an unfinished railroad line, while some 1,600 tired cavalrymen covered the flanks. But the situation remained critical, as some 3,500 Confederate troops pressed the attack, supported by about a dozen pieces of artillery. Confederate forces began to lap at the flanks of the Federal position with Brigadier General Joseph R. Davis' brigade working around the Yankee right as Brigadier General James J. Archer's brigade pressed the attack on his front and worked

A fine commander lost at Gettysburg, Major General John F. Reynolds. Reynolds was offered command of the **Army of the Potomac** *before the battle by President Lincoln, but turned him down.*

around the left. Reynolds ordered up the newly arrived Wisconsinites, Michiganders, and Indianans of the famed *Iron Brigade (1st/1st Division/I Corps)*, instructing them to "drive those fellows out of those woods." The veterans of the *2nd Wisconsin* came up in their distinctive big black hats, loading their muskets at the run. Shrugging off Confederate volleys, they plunged into the woods. Hitherto, many of the Confederate troops believed they were engaged against militia; they were soon disabused of this illusion. When the *2nd Wisconsin* attacked, some of Archer's men yelled, "Thar come the black hats! 'Tain't no militia! Its the *Army of the Potomac!*" as they fell back from the Federal left. Reynolds, who was observing the attack from horseback, was at that triumphant moment struck behind the right ear by a rifle ball and died almost instantly. Even as he toppled, to be carried off by some men of the *84th New York*, the balance of the *Iron Brigade* came up, delivering a series of three echeloned hammer blows to Archer's front and right. Archer's Brigade fell back to Herr Ridge, battered

The "Old Hero" of Gettysburg

Around noon on 1 July, as the men of *I Corps* were holding the line on McPherson's Ridge, Major Edward P. Halstead of the corps staff saw an elderly gentleman walking towards the ridge with a musket on his shoulder and a powder horn in his pocket. "Which way are the Rebels?" the old man asked, "Where are our troops?" Halstead told him that they were just ahead and that he would soon find them. The old man was John Burns, a resident of Gettysburg, and his blood was up. A few days earlier Rebel foragers had taken away his cows. He had sat in his rocking chair all morning listening to the fighting and decided to get into it. Burns, who was about 70, was no stranger to war, for he had served in the War of 1812, against Mexico, and as an Indian fighter. He soon took his place in the line alongside men young enough to be his grandsons. Burns fought with the *150th Pennsylvania*, the *7th Wisconsin*, and other regiments. The troops were at first amused by the little man who wore a swallow-tail coat with smooth brass buttons. They joked with him a bit. Seeing the ancient flintlock musket he carried they offered him a rifle. He waved it away, as he also did when given a cartridge box, saying "I'm not used to them new-fangled things." He fought well, like the veteran he was, and he accompanied the troops as they fell back to Seminary Ridge, in the retreat through Gettysburg, in the defense of Cemetery Hill, and in the defense of Cemetery Ridge on 3 July. He had high praise for the boys, telling Lieutenant Frank Haskell, "They fit terribly. The Rebs couldn't make anything of them fellers." The old man, who was wounded three times during the battle and at one point almost captured, soon became a popular figure. Lincoln himself asked to see him when he came to Gettysburg on the occasion of his famous address, and the two attended church together. Later a special Act of Congress granted him a pension of $8. He died in 1872, full of years and honor.

and exhausted, as its commander and hundreds of men were led to the rear as prisoners. The Federal left was saved. But the situation on the right grew worse.

Davis' Brigade had succeeded in flanking the Union position. Two of the regiments there were ordered back to Seminary Ridge. This exposed the remaining regiments of the *2nd Brigade*, and they shifted front, unmasking the Union center. Major General Abner Doubleday, who assumed command of *I Corps* at about 1100, threw in everything available,

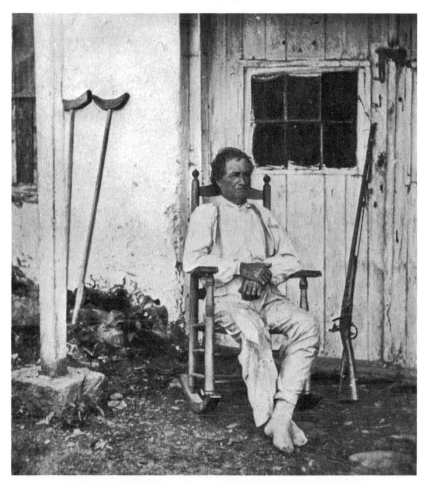

A veteran of the War of 1812, John Burns took up a musket once more to fight the Rebels who had invaded his neighborhood. He fought with Federal troops and was slightly wounded. After the battle, he was visited by President Lincoln and even awarded a pension for his brave services.

the brigade guard and the *6th Wisconsin*, 450 men in all. Posted on the left flank of the *Iron Brigade*, the reinforced regiment turned to the right and rushed northwards behind McPherson's Ridge. Just below the Chambersburg Pike the regiment came upon the exposed flank of the Confederate forces who were in pursuit of the two Union regiments falling back to Seminary Ridge. The Wisconsinites halted along a rail

Shortly before 1200, Confederate infantrymen of Colonel J.M. Brockenbrough's brigade of Heth's Division storm the stone barn of McPherson's farm, held by the Pennsylvania Bucktails.

fence and opened a voluminous fire into the Rebel flank. The Confederates took refuge in the unfinished railroad cut and opened fire in turn. A hot fight soon developed. Then the *95th* and *84th New York*, the regiments which the Confederates had been pursuing, formed up on the right flank of the *6th Wisconsin*. Then Lieutenant Colonel Rufus Dawes of the *6th Wisconsin* ordered an attack. In a rough line the Union troops vaulted the rail fence and drove straight at the enemy. Those who managed to get across the railroad cut began pouring a murderous fire in on the trapped Rebels. It was over in a few minutes, with many of the men of the 42nd Mississippi and 55th North Carolina becoming prisoners and all of the 2nd Mississippi; it was the first time a regiment of the Army of Northern Virginia surrendered. Davis' Brigade had been virtually destroyed; only the 11th Mississippi and a few hundred other men escaped back across McPherson's Ridge. The Union right was saved.

Ambrose P. Hill

A native of Culpeper, Virginia, Hill (1825-1865) graduated from West Point in 1847, 15th in a class which included Henry Heth, and entered the artillery. He served in Mexico, against the Seminoles, on the frontier, and in garrison before resigning as a first lieutenant on 1 March 1861. He entered Confederate service and was immediately commissioned colonel of the 13th Virginia, which he led in West Virginia and during the Bull Run campaign, though seeing little action. Appointed a brigadier general in early 1862, he commanded a brigade in the opening phases of the Peninsula campaign with such ability that he was promoted major general in May and given a division which he led with great skill, earning it the nickname "The Light Division" for the speed with which it executed all movements and maneuvers. Due to a personality clash between Hill and Longstreet, the division was transferred from the latter's command to "Stonewall" Jackson's Second Corps, with which it gained considerable distinction during the

Second Bull Run, Antietam, and Fredericksburg campaigns. At Chancellorsville the division formed the bulk of Jackson's striking force during his flank attack. When Jackson was mortally wounded, Hill assumed command of the corps until he was wounded.

In late May of 1863 Hill was appointed lieutenant general and given the newly created Third Corps, which he led into the Gettysburg campaign, where he demonstrated little brilliance, probably because he was so new at his task. Afterwards, however, he displayed some skill as a corps commander in all subsequent battles of the Army of Northern Virginia. During the final struggle for Petersburg in March of 1865, Hill rose from his sickbed to return to duty, giving an impressive performance until killed early in April, soon after the defense collapsed. Hill, who suffered from a variety of illnesses—some of them psychosomatic—was an excellent division commander but probably would not have been given a corps if he were not a Virginian.

By noon the situation on McPherson's Ridge had stabilized. The Union position was secure, at least for the moment, and Henry Heth had been given a serious drubbing. Firing died down and there was relative calm for two hours. Meanwhile, troops were on the march. One-armed Major General Oliver O. Howard arrived on the field at about 1130 with the advanced elements of his *XI Corps* and assumed command of the battle from Doubleday. He designated his *2nd Division* and corps artillery as a reserve, positioning them on Ceme-

A brilliant division commander, Ambrose Powell Hill took command of the newly created Third Corps before Gettysburg. His performance at the battle was merely adequate.

tery Hill, and began hurrying his forces forward as they arrived. At about noon he turned his corps over to Major General Carl Schurz. The balance of *I Corps* came up and was rushed to McPherson's Ridge, where Doubleday put his *3rd Division* in on the left and held his *2nd* in reserve. By 1400 hours two full corps were on hand plus Buford's cavalry, some 23,000 men and 60 pieces of artillery. Further back, other Union forces were beginning to move up.

Confederate troops were on the march too. Lieutenant General A.P. Hill, commanding Third Corps, decided to continue the action, apparently without clearing the matter with Lee. Heth brought up the balance of his division, putting some 7,000 men in front of McPherson's Ridge, while Major General William D. Pender's division was fast approaching with some 6,700 and Major General R.H. Anderson's 7,100 men were not far behind that. Other Confederate forces were on the move as well. On the morning of 1 July Lee had ordered Ewell's Second Corps to march on Gettysburg, though carefully instructing him to avoid bringing about a general action. By 1400 hours Ewell was some four miles north of Gettysburg with the divisions of Major General Jubal

Abner Doubleday

Doubleday (1819-1893) came from a family long prominent in upstate New York. His grandfather was a Revolutionary War veteran and his father had served in Congress. Abner himself served as a civil engineer for two years before entering West Point, from which he graduated in 1842, 24th in a class which also included Richard H. Anderson, Lafayette McLaws, and James Longstreet. Doubleday served in the artillery in Mexico and against the Seminoles, but had an otherwise uneventful career until April of 1861, when he was part of the tiny garrison of Fort Sumter. Thereafter he served in the Shenandoah Valley during Jackson's brilliant campaign in 1862 and at Second Bull Run, where he was a brigadier general of volunteers. He commanded a division in the *I Corps* at Antietam, Fredericksburg, and Chancellorsville, and led it into the Gettysburg campaign. On 1 July he assumed temporary command of the *I Corps* on the death of Major General John Reynolds and attempted to hold the Army of Northern Virginia north and west of the town. After the re-treat to Cemetery Hill, Doubleday returned to his division—Meade having reservations about his ability to lead a corps—which he led ably in the defense of Cemetery Ridge on 3 July. He held no further field commands during the war, serving in administrative posts in Washington. At the war's end he was named a major general of volunteers and colonel of the *35th Infantry*. His deliberate manner earned him the nickname "Old Forty-eight Hours."

As a field commander Doubleday was satisfactory but by no means brilliant, and Meade's decision to relieve him of the *I Corps* was probably correct. There is little beyond a very dubious tradition to identify Doubleday with the invention of baseball—the rules of which he is alleged to have written in 1839, when he would have been a 19-year-old cadet at West Point. A primitive form of the game was known as early as the Revolution. Both his brothers were in Federal service, one, Ulysses Doubleday, rose to brevet brigadier general of volunteers.

Early and Major General R.E. Rodes, about 15,000 men, while Major General Edward Johnson's 6,400 were following behind. But Longstreet's First Corps, some 21,000 of the best troops in the Army of Northern Virginia, was a long way off, and some elements had not yet taken to the roads, nor even been informed that a general action seemed imminent.

Colonel Thomas Devin, commanding Buford's cavalrymen to the north of Gettysburg, became aware of Ewell's approach

Major General Oliver O. Howard led the unlucky XI Corps *at Gettysburg. Despite the rout of his troops on 1 July, Howard somehow won the Thanks of Congress for his participation in the battle.*

at about 1230. Informed, Howard began moving his forces into position almost immediately. He posted the *1st* and *3rd Divisions* of *XI Corps* along a 2,500-yard front in the broad valley north of the town. At the same time, Doubleday extended his front about 1,000 yards northwards to the foot of Oak Hill by putting Major General John C. Robinson's *2nd Division* into the line. These movements were unwise. The two Union corps now lay at right angles to each other. Moreover, Doubleday's *I Corps* was overstretched on a front of more than 3,000 yards, and his left flank was still engaged with Heth's men. However, neither Howard nor Doubleday seem to have been concerned about this danger. Meanwhile Howard dispatched messages to Major General Daniel Sickles of *III Corps* and Major General Henry W. Slocum of *XII Corps*, informing them of the situation and calling for their help. He also reported to Meade. By this time Meade, a dozen miles southeast of the fighting, at Taneytown, was aware that a general action had begun. He also had been informed of the excellent work Reynolds had done in making a stand, and of the death of the latter, one of his closest friends. Meade ordered another trusted friend, Major General Winfield Scott Hancock of *II Corps*, to turn his command over to Brigadier General John Gibbon and assume command at the front. Still somewhat uncertain as to what the situation was, Meade instructed Hancock to report on whether the terrain and tactical situation at the Gettysburg position warranted a general battle, or whether the action should be broken off. Even as Hancock rode towards the front, the action grew more general.

Ewell's Second Corps struck in mid-afternoon, using Rodes' 8,000-man division. Rodes had concentrated his division on Oak Hill, but had placed one brigade in front of *XI Corps*, linking it with his main body by a line of infantry pickets, and held one brigade back as a reserve. Of his remaining three brigades, that of Brigadier General Alfred Iverson was to hit the Union front obliquely, while at its left that of Colonel E.A. O'Neal was to strike the Union front in the flank, and that of Brigadier General Junius Daniel was to cover the right of the two attacking brigades. At 1430 hours Rodes opened fire. Two batteries at the heavily wooded north end of Oak Hill took Doubleday's center, the *1st Division*, under enfilading fire. Brigadier General James Wadsworth, commanding, pulled his already battered right-hand brigade back diagonally across Seminary Ridge in an effort to link up with Robinson's *2nd Division* on the corps' right. This movement split the corps and Wadsworth's division as well, opening a 600-yard gap along the Chambersburg Pike. The left wing of the corps had 4,800 men in three brigades on McPherson's Ridge, while the right comprised Robinson's two brigades, plus Wadsworth's battered *2nd Brigade* and the *6th Wisconsin*, no more than 5,000 men. About 450 yards further to the right of *I Corps*, and roughly at right angles to it, was *XI Corps*. When Rodes' assault brigades moved out, *I Corps* seemed ripe for disaster. But even as the veteran Confederate regiments moved smartly to the assault, the

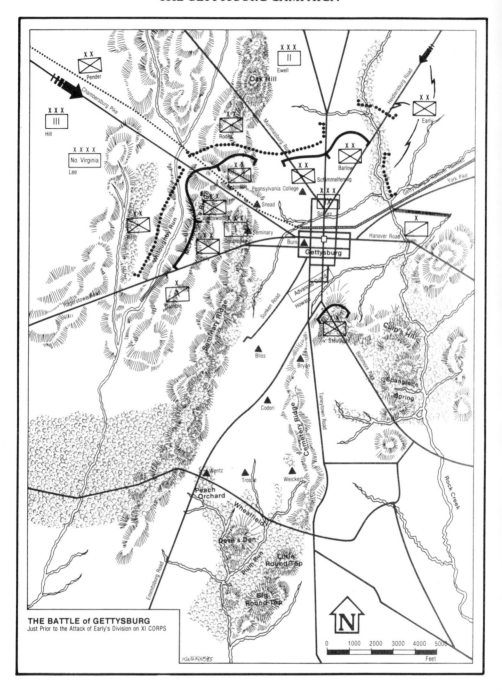

THE BATTLE of GETTYSBURG
Just Prior to the Attack of Early's Division on XI CORPS

attack began to falter. Rodes had failed to use his best brigade commanders, and neither Iverson nor O'Neal accompanied their men into action. Moreover, the latter attacked with only three regiments, rather than his full four, and did so on too narrow a front. Rapidly shifting his troops, Robinson took the attack of O'Neal's Alabamians frontally and threw the Rebel brigade back in disorder. This exposed Iverson's flank, but the latter kept coming anyway with some 1,400 North Carolinians in near perfect alignment. Under heavy artillery fire they began to falter, drifting off to the left, where they came under direct rifle fire, suffering perhaps 30 percent casualties in a few minutes. Then several Union regiments went over to the attack, storming forward to capture the survivors of the three regiments. The *Pennsylvania Bucktail Brigade (2nd/3rd Division/I Corps)* under Colonel Roy Stone covered the left end of the gap in Wadsworth's front, facing northwards and began pouring a heavy fire into Daniel's Brigade and remnants of Iverson's. Elements of *I Corps'* left wing contributed long range fire into the flank of the retreating Rebels. Within minutes it was over, as the remnants of Iverson's Brigade surrendered, to the distress of their commander, and as the survivors of O'Neal's shattered command drifted to the rear covered by Daniel's North Carolinians. Rodes' attack had failed, but he still had some fight left in him. Rapidly reorganizing the survivors, Rodes grouped them with his reserve, Brigadier General S.D. Ramseur's 1,000 strong brigade of North Carolinians, added 350 more of the 3rd Alabama, and held them ready for another attack. Meanwhile, Rodes' screen before *XI Corps* was having a busy afternoon.

Brigadier General George Doles' brigade of Georgians had done well during the initial stages of the fighting on the ridge to his right. Assisted by the 5th Alabama, Doles' 1,300 men had gradually forced Union Colonel Thomas Devin's cavalry pickets back. Infantry from the *3rd Division* of *XI Corps* came up, pushing Doles back. Counterattacking, he had run full into *XI Corps*, which promptly attacked on his right, intending to separate him from the balance of Rodes' Division.

Doles' shifted two of his regiments to meet the attack, but by about 1530 hours his situation was becoming critical and his left was collapsing. At that moment Ewell dropped the other shoe.

Jubal Early's division, about 5,500 veterans, had been getting into position north of Gettysburg while Rodes' troops were heavily engaged against both *I Corps* and *XI Corps*. Under the concealment of some thick woods, he had positioned a dozen heavy guns just east of the Heidlersburg Road, and deployed Brigadier General J.B. Gordon's brigade of 1,800 Georgians to the right of the road, Brigadier General Harry T. Hays' 1,300 Louisianans in the center, and Brigadier General Isaac Avery's somewhat fewer North Carolinians to its left, with Brigadier General William Smith's 800 Virginians behind Avery. By 1530 Early was ready. His artillery opened up, enfilading the front of *XI Corps*. Gordon's Brigade stepped off, moving to Doles' support at an easy pace so as not to tire the troops. Gordon soon located the right-flank brigade of *XI Corps*, some 1,100 men under Colonel Leopold von Gilsa, positioned with a stream to its front on the crest of a wooded hill just west of the road. Though strong in front, von Gilsa's position could easily be turned on its right, unhinging not merely the brigade, but all of Brigadier General Francis C. Barlow's *1st Division*. Demonstrating little tactical finesse, Gordon quickly threw his men against von Gilsa's front, while Doles' Brigade and the remnants of O'Neal's kept Barlow's other brigade, under Brigadier General Adelbert Ames, pinned down with rifle fire. Gordon's men were splendid, ignoring heavy fusillades to race some 900 yards across all obstacles, fording the stream and storming the hill. Despite the heroic efforts of the aptly named Lieutenant Bayard Wilkeson, just 19 years old, who kept his battery, *G, 4th Artillery*, in the front lines until mortally wounded, von Gilsa's brigade faltered. When Barlow fell wounded trying to rally the force, Ames took command, halted the rout and began forming a new line. He had barely begun when Early threw in the rest of his division. Hays and Avery struck Ames full in the right, shattering his improvised

At 1600 hours on 1 July, Lieutenant Bayard Wilkeson, just 19 years old, keeps G Battery, 4th Artillery at its work, covering the Union retreat in the face of Confederate Major General Jubal Early's attack which unhinged the Union lines in an action that would see the young officer mortally wounded.

line and unhinging the entire front of the division. This endangered the entire corps. By 1600 hours the corps had begun to fall back in considerable confusion. Schurz ordered it to rally on its reserve, Brigadier General Adolph von Steinwehr's *2nd Division*. Meanwhile Confederate arms were enjoying success all across the field.

With Doles' and O'Neal's Brigades already in action in support of Early, Rodes went over to the attack once more, pressing Daniel's Brigade against the weak southern end of *I Corps'* right. Heavy fighting developed and the issue remained in doubt for some minutes until Rodes' threw in Ramseur's reinforced brigade against the northern end of the corps. At almost the same time, Heth's Division reentered the lists, throwing four brigades against the left half of *I Corps*. The Yankees put up a terrific defense, inflicting heavy casualties on their assailants. Nevertheless, Heth's men gradually began to push the defenders back. At that moment, Lieuten-

Miss Shead's School for Girls

Miss Carrie Shead ran a school for young ladies in her family home on the Chambersburg Pike, between Seminary Ridge, and Gettysburg. On the afternoon of 1 July the battle of Gettysburg engulfed her home as Union forces began to fall back from Seminary Ridge. As the tired troops retreated towards Gettysburg, many of them took shelter in the Shead house. One such was Colonel Charles Wheelock of the *97th New York*.

Wheelock ran into the house, closely followed by several Confederate soldiers intent on taking him prisoner. He fled down to the basement, but the Rebels—and Miss Shead—followed him. A Confederate sergeant demanded that Wheelock give up his sword. The latter refused. Before the sergeant could take it, a second group of Rebels came down the stairs, herding some Yankee prisoners. After a moment's confusion the sergeant renewed his demand for Wheelock's sword. But it was gone, taken, Wheelock said, by one of the other Confederate soldiers. The sergeant went off angrily, deprived of his precious souvenir. Wheelock was later herded out of the house with several other prisoners, but managed to escape sometime thereafter. Several days later he returned to the Shead house to pay his compliments to Carrie and recover his sword, which she had hidden under her skirt.

The Shead house was crowded that day in July. Private Asa S. Hardman of the *3rd Indiana Cavalry* had fought all morning with the other troopers along Seminary Ridge. The Shead house provided him with temporary shelter as well. Hardman was also taken prisoner, but, unlike Colonel Wheelock, was unable to elude his captors. Eventually exchanged, Hardman also paid a return visit to the Sheads, so that he could marry Carrie's sister Louisa.

---

ant General A.P. Hill of Third Corps threw in four fresh brigades from Pender's Division. Attacking through Heth's brigades on a mile-wide front, Pender's men pushed *I Corps* back. Resisting stubbornly, the Union troops retired to Seminary Ridge. As they did, an heroic Bucktail color bearer turned briefly to wave "Old Glory" defiantly at the Rebels. He was cut down almost immediately. The Union troops reformed on Seminary Ridge, bolstered by 16 pieces of artillery firing almost hub-to-hub on a narrow front with Gamble's cavalrymen once more dismounting to cover their left flank. The Confederates still came on, though some regiments wavered. It was Brigadier General Abner Perrin's South Carolina brigade which proved the decisive element. Never faltering, it advanced with bayonets, smashing into the center of the newly formed Union line just south of the seminary which gave the ridge its name. Fierce fighting erupted around the buildings of the Lutheran Seminary, but the Rebels broke through and spread out to roll up the tattered ends of the Union line. Regiments began retreating to the rear through Gettysburg to meet those streaming back from *XI Corps* to the north, each taking a different road to safety. The Union attempt to defend Gettysburg had failed. But the defense had gained time. Moreover, though badly battered, with many missing—including both Barlow and brigade commander Brigadier General Alexander Schimmelfenning—neither corps had actually broken. Both had fallen back somewhat disorganized, but generally in relative good order. During the withdrawal they offered up stiff resistance in the town itself, particularly on the grounds of Pennsylvania College, the tall cupola of which provided an ideal observation post. The fighting in the town was very difficult for both sides, with wild melees occurring as regiments blundered into each other merely by taking a wrong turn. Union rear guards were generally effective. Union Captain Hubert Dilger positioned his *Battery I, 1st Ohio Light Artillery*, right in the Diamond, the main square of the town, sending round after round down the streets at the advancing Rebels. But by 1630 hours the situation had grown more desperate for the Union troops in the town. As Dilger's battery pulled back, the 1st South Carolina charged up, securing the square and pressing on to clear the town. The Confederates took perhaps 2,500 prisoners in the town itself, but they came away with few other trophies, securing only two or three battle flags and two pieces of artillery, one from each corps. Slender pickings indeed had the Union forces been genuinely smashed!

The Union retirement ended on Cemetery Hill. There, amid the tombstones and the monuments, von Steinwehr had established a substantial defensive position. From there his troops provided covering fire for the retreating regiments,

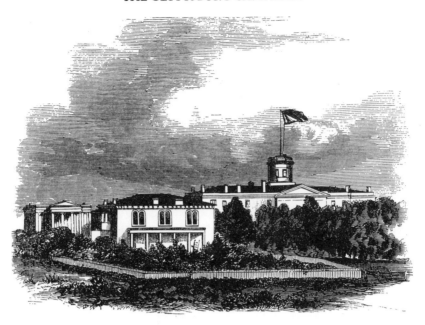

The seminary about which some of the fiercest fighting occurred on 1 July.

thereby checking the inadequate Confederate attempts to pursue. As additional troops came up they were put into the line, further bolstering the position. Meanwhile, Hancock had reached the field around 1615 hours. After a brief conference with Howard—who objected that he was senior and ought to be in command—Hancock assumed control of the Union forces. From atop Cemetery Hill, he discussed the situation with Howard, complimenting him on his choice of ground, finally saying, "Very well, sir, I select this as the battlefield," and then issuing orders to shift troops around. Characteristically, he met Doubleday's objections to the emplacement of Wadsworth's division on Culp's Hill with a burst of profanity that would have made a mule-skinner proud. The decisiveness of his actions, his bearing, and his determination were themselves inspirational and thus his very presence helped restore Federal morale. By 1800 hours Major General Henry W. Slocum's *XII Corps* began to arrive, taking up positions on the right, and bringing manpower up to about

The Barlow-Gordon Affair

The truth of this tale is difficult to ascertain, but it has been told and retold so many times that it has become a part of the tradition of the battle.

On the afternoon of 1 July, as the Union *XI Corps* began to fall back from its positions north of Gettysburg under heavy pressure from the Confederate Second Corps, Brigadier General Francis Barlow, commanding the corps' *1st Division*, was wounded in his left side. Unable to ride, he tried to walk off the field. As he did, he stumbled, collapsed, and was left for dead. Soon the tide of the advancing Confederate troops swept over him. As he lay on the ground, Barlow was spotted by Brigadier General John B. Gordon, who was leading the attack with his brigade of Georgians from Early's Division. Although he didn't know the seriously wounded 29-year-old Yankee brigadier general, Gordon stopped, gave Barlow what comfort he could, and had him carried to a Confederate field hospital. Speaking to the wounded man later, Gordon learned that Barlow's wife had accompanied the *Army of the Potomac,* much as his own had followed him to war with the Army of Northern Virginia. Gordon arranged for a safe conduct to be issued, and Barlow's wife was summoned to her husband's side. Barlow spent 10 months in a hospital before returning to duty, ending the war as a major general, as did Gordon.

After the war, each thought the other dead, Gordon because Barlow's wounds had seemed so grievous, and Barlow because a Brigadier General J. B. Gordon had been reported mortally wounded in May of 1864. In the post-war decades both men prospered; Barlow was several times elected secretary of state and attorney general of the state of New York, and was responsible for the prosecution of the infamous "Tweed Ring." and Gordon served his state as governor and for many years in the United States Senate. Then, about 20 years after the war, various notables were being introduced to each other at a dinner party in Washington.

When Senator Gordon and the secretary of state of New York were introduced, the senator politely inquired as to whether Barlow was related to that General Barlow who had died at Gettysburg. Barlow responded by saying that, he was, in fact the very man, having recovered from his wounds. With equal politeness Barlow inquired if Gordon was related to that General Gordon who had succored him on the battlefield but tragically died at Richmond in 1864. Gordon said that the officer in question was in fact his distant cousin, Brigadier General James Byron Gordon, and that he himself was the General Gordon who had been at Gettysburg. The two had a happy reunion, and remained friends thereafter.

Troops of the **Iron Brigade** *who fell on the first day of the battle.*

20,000 with 85 pieces of artillery. Major General Don Sickles' *III Corps* was near. The Federal position was solidifying. Slocum, who assumed command, concurred with Hancock's opinion that this new position could be held until Meade came up and decided whether to commit the balance of the *Army of the Potomac.* Hancock had already informed him that this was the place for battle.

The day's fighting had not gone well for the Union, but neither had it resulted in a disgraceful reverse. The Army of Northern Virginia had come off the victor in a hard fight against an inferior foe. But it was almost a pyrrhic victory. Confederate losses, while lower than the 12,000 killed, wounded, and missing suffered by the *Army of the Potomac*, had been substantial, perhaps 8,000 men. The results had not been worth the cost. To be sure, the advanced guard of the *Army of the Potomac* had been pushed back some miles, but it

Carnage of the first day of the battle. By the end of the battle both armies suffered a combined 50,000 casualties.

now occupied a position superior to that from which it had been ejected, and it was hourly receiving substantial reinforcements. Great tenacity and courage had been demonstrated, but little tactical finesse had been displayed, resulting in too many frontal assaults when flank attacks would have served. The troops had performed magnificently, repeatedly recovering the errors of their commanders by their courage and blood. But courage and blood can only do so much. Had the action been better handled, the Union would have suffered a far greater defeat at far less cost.

Union forces had done well, but could have done better. There was no question that the troops were skilled and determined, but some of their leaders were wanting. Had

Union forces been better led, they might have inflicted a modest defeat on the advancing Confederates before falling back to the positions in which they ended up anyway. They could, of course, have done worse, and an even greater reverse might resulted.

The day's fighting had set the stage for a more decisive clash on the morrow. And during the hot, humid, and cloudy night of 1 July, as generals laid their plans, tens of thousands of men slept in fields and barns and houses, while still more tens of thousands marched through the darkness.

Who Won the First Day?

The Confederacy won the first day's battle, though the leadership of the Army of Northern Virginia had been somewhat wanting. Lee made a number of serious errors by insisting on holding Anderson's Division in reserve throughout the fighting, when it might have turned a Union reverse into a disaster, and, more importantly, had not directed the battle. Instead, he suggested when he ought to have commanded. This was most notable late in the afternoon, when he gave Ewell the option of taking Cemetery Hill before dark, certainly one of the most unusual orders an army commander ever gave to one of his corps commanders. Lee also failed to coordinate the actions of his units, so that each division commander pretty much fought independently. Such coordination as did occur was largely fortuitous, as a result of individual commanders seizing the initiative.

Lee's subordinates had not done well either. To be sure, Hill had performed excellently against McPherson's Ridge in the afternoon, effectively coordinating the actions of two divisions. But Heth had completely bungled the opening action with Buford's cavalry, and gone on to do likewise in his initial efforts against the I Corps on McPherson's Ridge. Rodes and his brigade commanders had badly mishandled their attack on the I Corps, and Early had been slow in getting his division into action. Stuart's "raid"

severely hampered a more skillful conduct of the battle, for the Confederate generals had to operate without information which only his cavalry could have provided. Great tenacity and courage had been demonstrated, but little tactical finesse, resulting in too many frontal assaults when flank attacks would have served. Had the action been better handled, the Union would have suffered a far greater defeat at far less cost.

Although seriously defeated, the Union forces had done fairly well, but they could have done better. There was no question that the troops were skilled and determined, but some of their leaders were wanting. Buford's decision to dismount his troopers and fight north and west of Gettysburg was a good one, as was Reynolds' decision to support him. But it is doubtful that Reynolds intended to undertake anything but a delaying action, rather than seek a grand battle on two fronts, north and west of Gettysburg. For this Doubleday and Howard bear responsibility, particularly the latter, who put the XI Corps into the broad valley north of the town. Howard's selection of Cemetery Hill as a reserve position was noteworthy, however. Hancock arrived on the field too late to influence the course of the action. However, once there, he had done very well, by not taking counsel in fears, rapidly putting together a viable defense on Cemetery Ridge and

Culp's Hill so soon after a serious reverse and by urging that a stand be made in the new position. On the other hand, Slocum performed poorly because he failed to march to the sound of gunfire which his men heard shortly after noon whilst eating lunch at Two Taverns. Not six miles from the battle, he had missed an opportunity to turn the action into a draw. Lower level leadership, however, seems to have been skillful, despite Wadsworth's movements which split the front of the *I Corps*. Had Union forces been better led, they might have inflicted a modest defeat on the advancing Confederates. They could, of course, have been worse handled, and an even greater reverse might have resulted.

CHAPTER VI

The Struggle for the Flanks

2 July

Meade arrived on the field long before dawn on 2 July and immediately conferred with Howard, Sickles and Slocum at the caretaker's house of the local cemetery. Reviewing the situation, he decided to accept a general battle in the existing position, a decision which he had probably reached on the previous afternoon, for he had issued orders at that time directing the entire *Army of the Potomac* to march on Gettysburg. Then he took a quick look at the Union position on Cemetery Ridge, followed just before dawn by an extensive ride along the entire Federal line in the company of Howard and Brigadier General Henry Hunt, the chief of artillery. With the help of an engineer officer he produced a sketch map of the terrain, indicating on it the positions which he desired each corps to occupy. Copies of this were sent to each corps commander, who adjusted his lines accordingly. At about 0800 hours he sent Hunt to survey the lines once more for the purpose of selecting artillery emplacements.

The Union line was on ground averaging some 20 to 30 feet higher than that in front of it. Greatly resembling a fish hook, it was anchored on the right on heavily wooded, rocky Culp's Hill, 140 feet high, facing northwards and eastwards, fronting on some very broken ground and partially covered by Rock Creek. Culp's Hill was itself covered on the right by an extension of the line southward along Rock Creek for about

Henry J. Hunt

Born into an Army family in Detroit at a time when it was a frontier outpost, Hunt (1819-1889) was orphaned at the age of 10. He graduated from West Point 19th in the class of 1839 and entered the artillery, serving with distinction in Mexico (two brevets) and on garrison duty. In 1856 he was named a member of a board of three to review light artillery tactics. The report of this board was adopted by the Army in 1860 and became the basis for the artillery tactics of both sides during the Civil War. At the outbreak of the war he prepared the Harper's Ferry arsenal for "defense or destruction." Promoted major, he personally directed the battery which broke the Confederate pursuit after Bull Run. He briefly commanded the defenses of Washington and was then made colonel and aide-de-camp to McClellan, organizing the artillery of the *Army of the Potomac*. Hunt directed the artillery during the Peninsula campaign and was largely responsible for the Union victory at Malvern Hill.

Named a brigadier general of volunteers in September of 1862 Hunt directed the grand battery which opened the battle of Fredericksburg. When Major General Joseph Hooker took command of the *Army of the Potomac*, Hunt's authority was greatly reduced. The result was that the artillery was poorly handled and badly coordinated during the Chancellorsville campaign, whereupon Hunt was restored to a position of authority and conducted a thorough reorganization of the artillery, reducing the number of guns but greatly increasing flexibility. At Gettysburg he served as one of Meade's surrogates, and he was the first to point out the flawed deployment of *III Corps* on 2 July. On 3 July he directed the artillery barrage which helped shatter the Confederate grand assault. He remained as chief of artillery until June of 1864, when Lieutenant General Ulysses S. Grant named him to direct the siege of Petersburg, which he did with great technical skill. At the end of the war Hunt held brevet major generalships in both the volunteers and the Regular Army, but reverted to his permanent rank, lieutenant colonel.

In 1869 Hunt became colonel of the *5th Artillery*, a largely honorary post which he occupied until his retirement in 1883. He spent much of the latter part of his career on occupation duty in the South and later became governor of the Soldier's Home in Washington. Hunt was undoubtedly the finest artilleryman to come out of the war. He understood the important contributions which his arm could make, and fought long and hard to secure proper recognition for artillerymen, who were notoriously shortchanged in rank. His brother, Lewis C. Hunt, rose to brevet brigadier general during the war, serving in various theaters and administrative posts.

The best cannoneer of the **Army** of the Potomac, *Chief of Artillery* **Brigadier General Henry J. Hunt.**

1,400 yards. Slocum's *XII Corps* held the eastern side of the line here, with Wadsworth's division of *I Corps* on the northern face of the hill. About 450 yards northwest of Culp's Hill across some lower ground lay Cemetery Hill, some 600 yards south of Gettysburg. The northern end of Cemetery Ridge, the position of Howard's *XI Corps*, rises some 60 to 80 feet above its surroundings, with a panoramic view of everything to its northeast, north, and west, and has a fine, broad top suitable for several batteries of artillery. Cemetery Ridge stretches roughly southwestwards from there for about 1,500 yards, before turning roughly southeastwards for a similar distance. The ridge is not much more than 30 feet above the valley floor, and is gradually smaller as it trails off southwards. From its crest, the western slope of the ridge is deceptively easy, but the slope is considerably more formidable than it appears, rising in places at more than 30 degrees, with patches of broken ground and tangled vegetation. Situated here, facing roughly westwards, was the balance of *I Corps* under Major General John Newton (Meade having sent Doubleday back to his division) and, to its left, Hancock's newly arrived *II Corps*, 11,300 strong. To Hancock's left, Major

Daniel E. Sickles

A native of New York City, educated at New York University, Sickles (1819-1914) read law and jumped into politics, becoming a prominent leader of the Tammany Machine at an early age. He served variously as corporation counsel of the city of New York, at a diplomatic post in London, as a state senator, and as a member of the House of Representatives. In 1859, he shot and killed his wife's lover, Philip Barton Key (son of Francis Scott Key, composer of "The Star Spangled Banner") within sight of the White House. His defense, conducted by Edwin M. Stanton, who later became Lincoln's Secretary of War, was based on "temporary insanity," and he was the first person ever acquitted on those grounds. He subsequently forgave his wife, creating a further sensation.

At the outbreak of the Civil War Sickles raised the *Excelsior Brigade* in New York and became colonel of one of its five regiments, the *20th New York*. In September of 1861 he was made a brigadier general of volunteers and led a brigade of *III Corps* with commendable aggressiveness during the Peninsula campaign. Given a division, he led it at Antietam and Fredericksburg. Major General Joseph Hooker, a personal friend, then gave him his old *III Corps*, which he led at Chancellorsville, missing an opportunity to frustrate "Stonewall" Jackson's famed flank attack because he failed to act on his instincts.

At Gettysburg on 2 July the memory of this failure, combined with Sickles' natural aggressiveness and a lack of respect for Meade, led him to deploy his corps too far forward and over too wide a front, with the result that it was shattered by Lieutenant General James Longstreet's attack, Sickles himself los-

General Dan Sickles brought the 10,700 men of his *III Corps* into the line shortly after 0900, following a reprimand for dawdling from Meade. Sickles' left was at the end of Cemetery Ridge, just short of Little Round Top, a prominent, wooded, very rocky hill rising 170 feet above its surroundings. About 500 yards further south was Big Round Top, an extremely rugged, wooded, rocky hill rising over 300 feet above the valley floor. To the north and west of the Union lines the ground was relatively flat, open farmland with a scattering of buildings and occasional orchards, and stands of timber. The countryside in the vicinity of the Round Tops and to their south became increasingly rugged, with many trees, rocky outcroppings, and stray boulders. The entire Union

ing his right leg. He held no further field command, but performed various administrative functions for the balance of the war, while engaging in a disreputable smear campaign against Meade. After the war Sickles served on occupation duty in the Carolinas until 1869 when he retired from the Army as a Regular major general. He served for some time as minister to Spain, where he became the friend of Queen Isabella II; as he was a notorious womanizer and she was noted for her affairs, the possibility that they were more than friends is strong. The balance of his long life was spent in various political, civic, and business pursuits, with occasional trips to the Army's Medical Museum to visit his shattered bones, which he had donated as an exhibit. He also undertook an elaborate campaign which resulted in an award of the Medal of Honor for his conduct on 2 July 1863. As a member of the House he introduced the legislation which established the Gettysburg Battlefield Memorial Park on 12 February 1895, where he does not, however, have a monument. in 1912 he had to be removed from the chairmanship of the New York State Monuments Commission for peculation.

Sickles' political connections gained him a place in the army. His aggressiveness and courage in battle when many other officers were overly cautious secured for him high rank and ultimately a command for which he was unsuited. Fairly effective when leading a division under close supervision, he ought never to have been entrusted with a corps. Nevertheless, despite his flaws, there was much to admire in Sickles; as he was being carried to the rear with a shattered leg at Gettysburg he exhorted his men to fight harder.

---

front covered perhaps 6,000 yards. It was a good position, with broad fields of fire, many natural defenses, and short distances of communication. A thin line of skirmishers had been advanced from Cemetery Ridge and these covered the entire left of the Union line, giving additional security. With Meade's concurrence, Brigadier General Robert Tyler, commanding the *Artillery Reserve*, placed his guns in a central location in the rear, where they were no more than 2,200 yards from any point in the line. When Major General George Sykes' *V Corps* began to reach the field at about 0800 hours, after having marched some 50 miles in three days, Meade placed it next to the artillery. Meade had approximately 45,000 men and about 200 cannon holding the line behind breastworks and stone fences, plus 12,000 men and 124 pieces

The ever controversial political general Daniel E. Sickles. Sickles endangered the entire Army of the Potomac *on the second day of the battle by advancing his* III Corps *without permission thus exposing his flanks to enemy attack.*

of artillery in reserve. When Major General John Sedgwick's strong *VI Corps* with 13,600 and 30 guns marched up, Meade would have on hand some 70,000 men and over 350 guns.

Meade had several options. He briefly considered reinforcing his right and attacking Ewell's Second Corps. However, he gave up that notion when Brigadier General Gouverneur K. Warren, the chief of engineers, and Slocum reported that the ground in front of Culp's Hill was unsuitable for an attack. As a result, he had second thoughts about attempting an offensive move. He had to expect that the enemy would not remain inactive. Lee was a resourceful, wily, aggressive foe, and it would be best to see what he was up to before committing the army. The natural strength of the position further encouraged Meade to adopt a defensive stance. So, while Union and Confederate troops exchanged rifle and artillery fire at a distance, Meade spent his time getting his troops into order, issuing instructions, and surveying his lines. He left the details of the deployment of each corps to its commander, merely indicating in a general way the positions he wished covered. This generally worked, for both officers and men were well trained and experienced. Most of the corps commanders understood that their corps were but individual parts of the whole and made their dispositions accordingly, being careful to establish contact with the neighboring corps on each flank. Unfortunately Sickles of III Corps was an amateur soldier. Though commendably aggressive and a good battlefield commander, he had a tendency to be careless. Initially he had positioned his troops properly to the left of Hancock's II Corps. But in the afternoon, he pushed forward, perhaps hoping to feel out the enemy. As a result, his two divisions ended up at roughly a 45 degree angle to each other, forming a triangular salient with its apex nearly 1,500 yards in advance of the Cemetery Ridge position and its base over 2,000 yards wide. As deployed, Sickles' right consisted of Brigadier General A. A. Humphreys' 2nd Division with three brigades along the Emmitsburg Road for some 650 yards, and his right hanging about the same distance in front of Hancock's left brigade, which he partially masked. On Humphreys' right was Brigadier General Charles Graham's brigade of Major General David B. Birney's 1st Division. Graham's men would form the apex of the salient, with half its troops holding a front of about 450 yards along the road, and the other half at an immediate right angle, holding a similar front along a country lane running around John Sherfy's 12-acre peach orchard. About 400 yards to Graham's left the balance of Birney's division, two brigades, occupied a very strong position stretching about 1,000 yards roughly east-to-west in front of some wheatfields and an almost lunar landscape of rocky outcrops and boulders called the Devil's Den. The left was at Plum Run, a small stream running through a modest, boggy valley just west of the Round Tops. As Henry Hunt observed, the deployment of Sickles' units was not too bad, taken individually. The brigades were generally in good positions, well covered to their front, with excellent fields of fire. Unfortunately, taken as a whole, the situation of the corps was not good. Its line was too long for the manpower available, with only 1.6 rifles per yard of front. Then too, there was the awkward angle in the middle of Graham's brigade. What was worse, the entire corps merely hung in the air, with no flank protection and no liaison with

James Longstreet

A native of South Carolina, Longstreet (1821-1904) graduated from West Point 54th out of 62 in the class of 1842. He served in the infantry against the Seminoles, with distinction in Mexico (one wound, two brevets), on the frontier, and in garrison until resigning on 1 June 1861 as a paymaster major. He immediately sought a similar appointment in Confederate service but was instead made a brigadier general and commanded a brigade at Bull Run the following month. That October he was promoted major general and given a division, which he led with distinction in the opening phases of the Peninsula campaign in early 1862. He performed poorly when given a higher command shortly thereafter, but redeemed himself during the Seven Days Battles and was ever after in Lee's confidence, commanding First Corps. He led his corps during the Second Bull Run and Antietam campaigns, was promoted lieutenant general and commanded his corps at Fredericksburg. In early 1862 the corps served briefly on the Carolina coasts, returning to Virginia in time for the Gettysburg campaign, about the wisdom of which he had serious reservations. During the battle of Gettysburg, Longstreet's Corps bore the burden of the fighting, particular on 2 and 3 July. Longstreet was notably dissatisfied with Lee's offensive plans for both days, and his doubts have tended to be confirmed in retrospect.

After Gettysburg, Longstreet took his corps west in September of 1863 to support Braxton Bragg's operations and he was largely responsible for the Confederate victory at Chickamauga. Returning to Virginia in early 1864, Longstreet was seriously wounded by his own men at the Wilderness in May, and did not return to duty until October. He then led his corps in all the succeeding battles of the Army of Northern Virginia until Appomattox, where he was the only general officer to urge continued resistance if the terms offered by Grant were dishonorable.

After the war Longstreet settled in New Orleans, where he became a Republican and a supporter of Reconstruction, once commanding black militiamen against a rioting white mob. This resulted in a vicious campaign to smear his reputation, led most notably by Jubal Early and William Nelson Pendleton, who attempted to place on him all blame for the reverse at Gettysburg, despite the fact that he retained Lee's confidence to the end. He afterwards was made minister to Turkey by President Grant, who became a close friend, wrote his memoirs, and served in a variety of other government posts. Nicknamed "Old Pete" by the troops and "Old War Horse" by Lee, Longstreet was an excellent corps commander with a good eye for ground and a fine tactician, whose reputation was long in recovering from the attacks of his enemies.

One of Lee's most reliable corps commanders, **Lieutenant General James Longstreet.** *Throughout the battle, Longstreet argued with his commander over the wisdom of attacking the* Army of the Potomac *at Gettysburg.*

the rest of the army. Moreover, Sickles had neglected to inform Hancock, to his right and rear, of his deployment. In short, the left flank of the *Army of the Potomac* was exposed. A determined attack could easily crush *III Corps* and then proceed to roll up the entire Union line.

Lee was determined to attack on 2 July. Generally satisfied with the outcome of Wednesday's fighting—when some 5,000 prisoners had been taken—he believed he would again do well on Thursday, a belief generally shared by all of the officers and men of the Army of Northern Virginia. The problem was how to go about it. Longstreet, Lee's closest and most able subordinate, disliked the looks of the Union positions. Conferring with Lee on the evening of 1 July, he pointed out the significant natural advantages of the lines held by the *Army of the Potomac* and the fact that, in the absence of adequate reconnaissance, they could not be certain as to the strength of the Union forces. He then suggested a broad sweeping movement on the Confederate right. A strategic maneuver rather than a tactical one, such a move would put the Army of Northern Virginia between Meade and Washington, thereby forcing him to abandon his present excellent position. More importantly, the maneuver would permit the Army of Northern Virginia to stand on the defensive, for Meade would be forced to attack in order to restore his communications with Washington. And the Army of Northern Virginia had never yet done badly in a defensive fight. Lee seems to have considered Longstreet's proposal, but demurred. This decision has at times been attributed to exhaustion, illness or even blood lust, but it was by no means irrational.

Though Longstreet's proposal was attractive, it had several drawbacks. The absence of Stuart made dangerous any spectacular movements in the face of the enemy, particularly as Lee had no knowledge of the locations of Meade's corps beyond the two he had already encountered. Nor could the movement be made immediately, for only about 70 percent of the Army of Northern Virginia would be on hand on the morning of 2 July; delaying the maneuver to await the arrival of the balance of the army might give the day to Meade. Then too, the morale of the army might suffer if it appeared to be abandoning its hard-won gains and moving south once more. When Longstreet urged that they at least stand on the defensive in their present position, Lee also disagreed. The army could not remain for long inactive in such close proximity to the *Army of the Potomac* in the heart of Pennsylvania. Meade might choose not to attack, but rather to essay a strategic maneuver of his own. Or he might remain in position, thereby pinning the Army of Northern Virginia, while Union troops poured into central Pennsylvania from all directions to reinforce him, or, worse, develop a strategic trap by moving into the Shenandoah Valley and blocking Lee's line of retreat. The Army of Northern Virginia would have about 60,000 mostly fresh troops and some 200 pieces of artillery at hand on 2 July to face a somewhat superior, but already battered foe. Lee had Meade pinned, and he intended to hit him.

Lee proposed to attack on the Union flanks. Over on his left Ewell's Second Corps would attack the Federal right in the morning, followed soon after by Longstreet's First Corps, which would assault Cemetery Ridge from the Confederate right. Ewell's attack would be primarily a diversion, designed to keep Union forces in place. Longstreet's would be the main blow. It took Lee a while to come up with this plan, and he changed it several times in the face of objections from his generals. In the end he had to insist on it, despite the fact that Ewell believed his corps should stand on the defensive and leave the attacking to Longstreet, while the latter was in favor of a defensive posture for the entire army. Lee used his authority as commanding officer to overrule their objections, and they set about preparing for the day's action.

Lee ought to have tidied up his lines a bit. The Army of Northern Virginia held a concave front stretching nearly six miles. Ewell's Second Corps, on the Confederate left, had one division covering its left flank and holding a line running about 1,200 yards on a roughly northwest-southeast axis, and two others holding about 1,800 yards on an east-west axis, running right through the heart of Gettysburg. Most of Ewell's units were between 500 and 1,200 yards from the Union forces holding Culp's Hill and Cemetery Hill. With a bend in his front, a few low hills to the left, Gettysburg in his center, and open fields on his right, Ewell's lines were not particularly suited to the defense, but Lee appears to have ignored the matter. In Lee's center, to the right of Second Corps and at a right angle to it, was Hill's Third Corps, with one division in reserve and two holding about 3,000 yards of front along Seminary Ridge, a useful defensive line, running roughly parallel to the Union forces along Cemetery Ridge some 1,200 yards to the east. To Hill's right was Longstreet's First Corps, which on the morning of 2 July had yet to arrive on the field, but which, when deployed, would hold a front of about 2,500 yards right before the ground into which Sickles would intrude his *III Corps* later that morning. With perhaps 40,000 troops in line, Lee would have about four men per yard of front, which was not bad. But the curvature of his front left much to be desired for Ewell's Second Corps was poorly situated being too far from the rest of the army. It would be difficult to shift troops rapidly to its support should the force become endangered. Ideally, it ought to have been pulled back towards the right, thereby greatly shortening Lee's front and improving his communications. As it was, Lee had suggested doing this during the heated staff conferences on the night of 1-2 July held in the Adams County Prison on High Street in Gettysburg, but had allowed himself to be talked out of it by Ewell and his very persuasive subordinate Jubal Early. He then made a few suggestions, including one that Ewell occupy Culp's Hill, which had been reported as free of enemy troops. Then he left his commanders to prepare the details of the coming day's action.

During the night the Army of Northern Virginia got ready for battle. This was a difficult task for Longstreet as his corps was not yet on the field. Major General Lafayette McLaws' division camped nearby at about 0100 after eight hours on the march. Longstreet's artillery battalion and Major General John B. Hood's division marched through the night to reach the vicinity of the battlefield during the pre-dawn hours. Before Longstreet could do anything, these units would have to be brought to the front, the artillery would have to be sited and the troops would have to be fed and rested. Aware of these difficulties, Lee altered his plans somewhat. Longstreet would attack as soon as all preparations were made, while Ewell was ordered to defer his attack until he heard the sound of Longstreet's artillery. For Ewell these instructions were welcome since during the pre-dawn hours a disturbing complication had developed. Acting in accordance with Lee's instructions, he had sent Major General Edward Johnson to occupy Culp's Hill. Advancing on the position, Johnson had found it swarming with Federal troops. Moreover, a captured dispatch revealed that the Union *XII Corps* was already on the field and that two divisions of their *V Corps* were not four miles to his left and rear. Passing word on to Ewell, Johnson held his position. Meanwhile Lee was still rethinking his plans, so that even as Ewell learned of the potential danger to his flanks, he received new orders from Lee. These granted Ewell virtually complete freedom of action, for they instructed him to make a demonstration against the Federal right when he heard the opening rounds of Longstreet's attack, to deliver a general assault only if he felt the opportunity right. Ewell shifted a few brigades to cover his left and waited for Longstreet.

During the oppressively humid morning of 2 July, preparations for the Confederate attack were well in hand. The last of Longstreet's troops, his artillery, came up at about 0800 and immediately took to their blankets for a couple of hours rest. Lee was everywhere, visiting in turn each corps. The delay was fortuitous. There was some uncertainty as to the nature of the ground over which Longstreet would have to attack and as to the location of the Union lines. Several officers sent out on reconnaissance had brought back useful but inconclusive information. Shortly after dawn, Lee decided to dispatch Captain S.R. Johnston, an engineer from his staff, on a proper reconnaissance. With a small group of companions, Johnston explored the ground between Seminary Ridge and Cemetery Ridge, quietly penetrating the Union picket lines and scouting out the entire area up to the woods at the southern end of Cemetery Ridge. He even climbed the steep slope of Little Round Top to observe that there were no concentrations of Union troops anywhere to be seen south of Cemetery Ridge. By coincidence, Johnston's party had missed the movement of the leading elements of Sickles' *III Corps* into the line south of Cemetery Ridge by no more than a few minutes, their location being concealed by Cemetery Ridge and the woods themselves. At about 0900 hours Johnston returned to Lee, delivering his report. Lee was pleased, for Johnston's report suggested that there were no Union troops directly in front of Longstreet's Corps. He began to discuss the tactical details of the attack with Longstreet and McLaws, whose division was to spearhead the attack, while Hill looked on. Lee wanted the attack delivered along the Emmitsburg Road to drive northeastwards into the Union lines on Cemetery Ridge. Longstreet disagreed, wanting the attack to be made directly eastwards across ground that Johnston said was bare of Union forces. Lee overruled him, but the latter persisted and an unpleasant scene developed between the two old friends.

John B. Hood

A native of Kentucky, Hood (1831-1879) graduated from West Point 44th in the class of 1853 and entered the infantry. He served in garrison and on the frontier, where he was wounded, before resigning from the Army as a first lieutenant in April of 1861. He immediately entered Confederate service in the same rank. He held a variety of minor commands with some distinction until the Peninsula campaign, when he was named brigadier general. He was given a brigade of Texans which he led with skill and vim at Gaines' Mill, Second Bull Run, and Antietam. In October of 1862 he was named major general and given a division in Longstreet's First Corps. He led his division at Fredericksburg, in the Carolinas, and at Gettysburg. Hood's Division made the principal Confederate attack on 2 July, shattering the Union *III Corps*, during which his left arm was crippled.

Hood went west with Longstreet later in 1863, briefly commanding the corps and several divisions of the Army of Tennessee at Chickamauga, where he lost his right leg. Remaining in the west, he was given a corps in Gen. Joseph Johnston's Army of Tennessee and promoted lieutenant general. He led his corps with commendable aggressiveness in the opening stages of the Atlanta campaign, during which Johnston's clever maneuvering kept Federal gains to a minimum. In July Hood was made an acting general and superseded Johnston in command, an action dictated by Jefferson Davis' dissatisfaction over Johnston's refusal to fight. Hood did fight, but Sherman won all the battles, securing Atlanta, and ultimately launching his famed "March to the Sea." In a desperate move, Hood launched his army into Tennessee, hoping to divert Sherman's attention, but was himself defeated at Nashville, becoming the only commander in the war to lose his entire army in the field. Relieved at his own request, he reverted to his permanent rank but saw no further service.

After the war Hood became a merchant in New Orleans, married, and had 11 children in 10 years. The yellow fever epidemic of 1878 caused a financial crisis which wiped out his fortune and in the following year carried off him, his wife, and one of their children. A big, aggressive man who loved combat, Hood was not a subtle commander and had an unfortunate tendency to blame his subordinates for his own errors. His skill as a commander deteriorated markedly after the loss of his leg.

In the end, of course, Lee had his way. Longstreet's Corps would attack in two stages: there would be an initial attack into the area west of the Round Tops to secure artillery

The hard-fighting Major General John Bell Hood. Hood was severely wounded during his division's attack against the Federal left.

positions, followed by an advance along the Emmitsburg Road designed to drive in the Union left flank at the southern end of Cemetery Ridge. By about 1000 hours, Longstreet was issuing orders for his artillery to get into position to support the attack. Lee then rode off to see Ewell. At this point Longstreet dawdled, awaiting the arrival of Major General George E. Pickett's division and a stray brigade of Hood's Division. He was still concerned about the wisdom of the proposed attack and did not believe he had enough strength to bring it off. When Lee discovered this delay upon his return to Longstreet's headquarters at about 1100 hours, he demanded that the attack be made as soon as all of Hood's Division was at hand. At noon Longstreet's Corps finally began to deploy, moving somewhat to the right in order to get into proper position from which to launch the attack. The movement was badly conducted despite the fact that Captain Johnston was assigned to guide McLaws' Division along a concealed route. Unfortunately, the ground over which the division was to move was not that which Johnston had spent much of the morning exploring. The result was that the movement became confused with lengthy countermarches and tiring cross-country hikes, and with some of Hood's men getting entangled in McLaws' ranks. It appears that Federal signalmen on Little Round Top observed at least part of the movement at about 1330 hours, which was precisely what Lee had hoped to avoid. The troops only began to occupy their assigned positions at about 1500. And at about that time, McLaws' advanced guard, the brigade of Brigadier General Joseph B. Kershaw, encountered strong Union forces in the vicinity of the Peach Orchard. Unknowingly Longstreet had run into Sickles' exposed *III Corps*. McLaws halted his division and prepared for an attack. While Longstreet insisted that the troops before him were insignificant, McLaws countered by arguing that he was facing a superior foe. Convinced, Longstreet called up Hood's Division and ordered a full-scale attack.

The attack would be made as soon as the troops were in position. McLaws' Division was on the left, with two brigades up and two following behind, on a front of roughly 1,000 yards, not 500 yards from the apex of Sickles' salient at the Peach Orchard. To McLaws' right, and at roughly right angles to it, was Hood's Division. Hood's right rested just short of Big Round Top and he had four brigades in line on a front of about 1,400 yards, between 500 and 800 yards from Sickles' left flank. Longstreet committed 54 pieces of artillery under the able Colonel Edward P. Alexander, who sited them on a 1,000-yard front some 1,800 yards west and southwest of the Peach Orchard. By 1600 hours all was in readiness for Longstreet to fall on Sickles' *III Corps*.

By mid-afternoon the deployment of Sickles' corps had begun to disturb Meade. He ordered *V Corps* to move to Sickles' left, replacing it in the reserve by the newly arriving *VI Corps*. Then, prompted by Henry Hunt's report, he rode over to examine Sickles' lines, arriving at *III Corps* a little before 1600 hours. He was appalled by what he found. Sickles' advanced position threatened to unhinge the entire *Army of the Potomac*. The corps commander apologetically

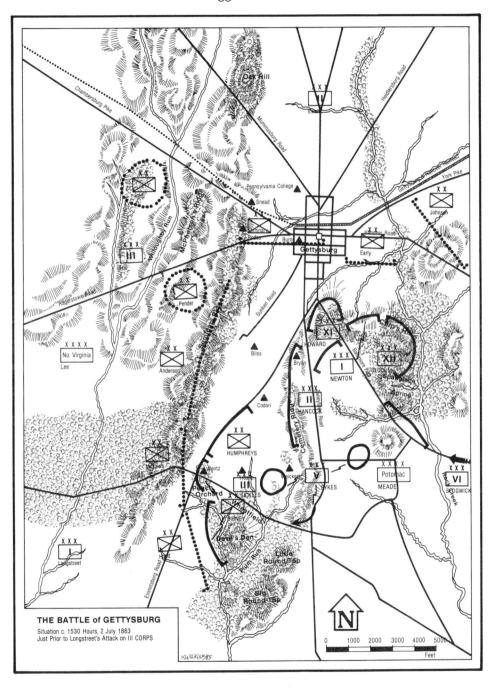

THE BATTLE of GETTYSBURG

Situation c. 1530 Hours, 2 July 1863
Just Prior to Longstreet's Attack on III CORPS

Lafayette McLaws

Born in Augusta, Georgia, McLaws (1821-1897) graduated from West Point 48th in the class of 1842, along with James Longstreet. He served in the infantry in Mexico, in garrison, on the frontier, and in the "Mormon War," rising to captain before resigning from the Army on 10 May 1861. He entered Confederate service as a major and was soon commissioned colonel of the 10th Georgia and was shortly afterwards made a brigadier general, commanding a brigade in the opening stages of the Peninsula campaign. He was soon promoted major general and given a division under "Stonewall" Jackson, which he led in the Shenandoah Valley, and then under Longstreet in the Antietam and Fredericksburg campaigns. McLaws' particularly distinguished himself in the defense of Marye's Heights during the Chancellorsville campaign, when his division held back an entire corps. At Gettysburg his division performed well on 2 July. During Longstreet's operations in the West in late 1863, McLaws was relieved by his commander for lack of cooperation, but he was subsequently cleared of all charges. His relations with Longstreet were stormy thereafter and he was sent to an administrative command in the Carolinas, and later directed the defense of Savannah. After the war he was in the insurance business, and served as collector of internal revenue and postmaster at Savannah.

McLaws was an excellent division commander when clearly told what to do and closely supervised by his superiors, having no talent for independent command. A loyal officer, despite his personal differences with Longstreet, he supported the latter against the attempts of Early and Pendleton to smear his reputation after the war.

offered to withdraw his corps to Cemetery Ridge, but Meade pointed out that it was too late for that. He was right. Within minutes Rebel artillery began to open up on Sickles' front. Meade directed Hunt to bring about 50 guns from the *Artillery Reserve* to the support of Sickles and ordered *V Corps* to come up as quickly as possible. Throughout the morning and early afternoon there had been little skirmishes and brief artillery exchanges all along the front. Now it was time for more serious fighting.

Alexander's artillery opened up at 1600 hours, taking the Peach Orchard under heavy fire. Within minutes Confederate infantry began to advance. As planned, Hood's men were to

Major General Lafayette McLaws, a division commander in Long-street's Corps.

go in first, followed within the hour by those of McLaws. Hood's troops advanced at a fast pace, with their front well covered by skirmishers. The right hand brigades under Brigadier General Evander Law, with 1,900 men, and Brigadier General Jerome Bonaparte Robertson, with 1,700, drove somewhat westwards, towards Big Round Top. Union artillery opened up. Union skirmishers from two regiments of

A wartime sketch of the fighting for Devil's Den, showing Hood's Confederate infantry converging on the Union position from two directions, the left and the background.

sharpshooters added their fire, but the bulk of the Federal infantry waited. Despite the Yankee artillery and musket fire and broken ground, the Confederate ranks maintained their cohesion and threw back the thin line of skirmishers. Reaching the gorge of Plum Run on the western side of the foot of the Round Tops, Law's Alabamians swung to their right, some of the men advancing over rugged Big Round Top while the balance moved up the creek and on to the slopes of Little Round Top to get around Sickles' flank at Devil's Den. Robertson's men, mostly Texans with a regiment of Arkansans over on Law's left, stormed right into the left flank of Brigadier General John Ward's brigade, 2,200 men of Sickles' *1st Division*, holding the ground in front of Devil's Den. Seeing the battle beginning to develop nicely on his right, Hood ordered his other two brigades, Brigadier General Henry Benning's with 1,400 and Brigadier General George T. Anderson's with 1,900, to support Law and Robertson. Ward's men held their fire until the Rebels were within 300 yards, whereupon they delivered a brigade-sized volley directly into the charging Confederates. The attackers faltered momentarily. Hastening to reload, Ward's men got off yet another volley, further confusing the enemy. Then Ward counterattacked, driving the Confederates back and securing a stone wall forward of his original position. The Confederates came on again, driving Ward back. Once more he counterattacked. Some of the fiercest fighting of the war developed along the entire front of Birney's *1st Division*, as Hood's brigades came into action one after another. Places of no importance became hotly contested battle grounds where men fought and died by the score: the Peach Orchard, the Wheatfield, Devil's Den. Sickles' veterans fought hard and well and held their ground. At one point, an officer protested to Colonel A. Van Horne Ellis of the *124th New York* that by mounting in the face of a Confederate attack on the second day at Gettysburg, he and Major James Cromwell were exposing themselves unnecessarily. The colonel replied, "The men must see us this day," and fell in action shortly afterwards. The regiment held. But despite such heroism, the position was inherently flawed, and the flaws began to take their toll.

Over on Sickles' left was Little Round Top, rising high above the surrounding countryside. Earlier, before the crisis erupted in front of *III Corps*, Meade had instructed his chief engineer and good friend Brigadier General Gouverneur K. Warren to investigate the situation on Little Round Top. Riding swiftly there with Lieutenant Washington A. Roebling and a few other aides, Warren reached Little Round Top shortly after Longstreet's attack on Sickles got underway.

Gouverneur K. Warren

Born in New York at Cold Spring, Warren (1830-1882) graduated from nearby West Point second in the class of 1850 and thereafter served in the Corps of Topographical Engineers and as an instructor of mathematics at the Academy. At the start of the war he was lieutenant colonel of the *5th New York*, seeing action at Big Bethel in June of 1861. He soon became colonel of the regiment and led it in the Peninsula campaign, where he acquired a wound and a brigade in *V Corps*, which he led with distinction at Second Bull Run and Antietam. He was promoted brigadier general of volunteers in September of 1862 and led his brigade at Fredericksburg. That spring he joined the staff of the *Army of the Potomac*, rising to major general in May and becoming chief of engineers in June.

It was during the Gettysburg campaign that Warren truly distinguished himself. Right at the start, when Hooker was still uncertain as to Lee's intentions, Warren concluded that he was moving into Pennsylvania and procured the necessary maps. At Gettysburg the most important of his many services was his improvised defense of Little Round Top on the afternoon of 2 July. After Gettysburg he assumed command of *II Corps* during the convalescence of Major General Winfield Scott Hancock, and the following spring was given *V Corps*, which he ably led through all the battles of the campaign of 1864 up to Five Forks, when he was relieved by Major General Philip Sheridan, with the consent of Lieutenant General U.S. Grant. Though no reason was given for this at the time, Warren's relief appears to have been due to a personality clash with Sheridan, who later stated that Warren had been lethargic in his movements during the battle. This effectively ruined Warren's career. Thereafter he held various administrative posts.

Remaining in the Army after the war, Warren did not rise to lieutenant colonel of engineers until 1879, when after years of requests on his part a court of inquiry was held which exonerated him, concluding that there had been no grounds for his relief. Aside from the published verdict of the court and his belated promotion, Warren received nothing. A good man, and a fine general, Warren deserved better for his distinguished and honorable services to the Republic.

There he was stunned to discover that, save for a few signalmen who were packing up to leave, the hill was bare of Union troops. Warren ordered the signalmen to get back to work as conspicuously as possible, and then, knowing that Major General George Sykes' *V Corps* was approaching, he

Strong Vincent

Born in Pennsylvania, Vincent (1837-1863) graduated from Harvard in 1859, studied law, and was admitted to the bar in 1861. On the outbreak of the war he joined the *Erie Regiment*, a three-months militia outfit, as a lieutenant and adjutant. Upon mustering out, he became lieutenant colonel of the *83rd Pennsylvania*, with which he served during the opening stages of the Peninsula campaign. Ill for some time thereafter, he nevertheless was named colonel of the regiment later the in campaign, but did not rejoin the unit until Fredericksburg, where it was heavily engaged. He led his regiment at Chancellorsville and was then given a brigade in the *V Corps*. It was Vincent's brigade, and notably Joshua Chamberlain's *20th Maine*, which held Little Round Top on the afternoon of 2 July, anchoring the Union left. During the fight Vincent was mortally wounded trying to rally the *16th Michigan*, and died on 7 July.

dispatched two messages, one to Meade recommending that a division be brought up immediately and one to Sickles in an effort to pry some troops from him. Sickles, hard pressed, refused, but Meade overruled him and ordered A. A. Humphreys' division, already under enemy fire, but not yet deployed on the Emmitsburg Road, to reverse its march. Within minutes Meade, learning that *V Corps* was already on the field with one division in Sickles' rear, canceled the order. Humphreys countermarched once more. Warren, meanwhile, was impatiently watching the progress of the battle in front of Ward's brigade. He rode out to locate *V Corps* and within a few minutes found Sykes, who was examining the ground in the rear of *III Corps*. Warren apprised Sykes of the situation on Little Round Top and the latter immediately dispatched an order to Brigadier General James Barnes, commanding his *1st Division*, to place a brigade on the hill. Unfortunately, the courier could not locate Barnes. But he did run into 26-year-old Colonel Strong Vincent commanding Barnes' *3rd Brigade* of 1,300 men. Vincent pried the message from him. Recognizing its importance, on his own initiative and in complete disregard for his orders, Vincent immediately got his brigade on the march to Little Round Top. The brigade swiftly covered the half-mile to Little Round Top, racing over fields

William C. Oates

A native Alabamian, as a young man Oates (1835-1910) earned his living as a house painter, then read law. He edited a newspaper and practiced law for several years while dabbling in local politics. On the outbreak of the Civil War he recruited a company which became part of the 15th Alabama, of which he was soon named colonel. During the war he led his regiment ably and courageously in 27 engagements in the Shenandoah Valley, Northern Virginia, Tennessee, and around Richmond, but most notably at Gettysburg, where his attempt to storm Little Round Top was frustrated by Joshua Chamberlain's *20th Maine*. In August of 1864 he lost his right arm in the fighting around Richmond, but returned to duty and remained with his regiment until Appomattox.

After the war, Oates resumed his profession in Alabama, becoming quite prosperous, and entered state politics. He served in the state legislature and as a delegate to a state constitutional convention. In 1880 he entered the House of Representatives, where he became an outspoken segregationist and nativist. He served in the House until 1894 and for a time held the equivalent of a filibuster championship, speaking for eight days. In 1894 he became governor of Alabama. Four years later he returned to military service when President William McKinley made him a brigadier general of volunteers, though he saw no action during the Spanish-American War. He afterwards practiced his profession until his death.

and pastures at the "double quick," crossing Plum Run and, under fire, sweeping up the west side of the hill and on to its wooded rugged crest. Colonel Joshua L. Chamberlain's *20th Maine* was in the lead and as it streamed down the southern face, Vincent placed it on the extreme left. Then he went off to direct the deployment of his other regiments to its right as they came up. Before he left, his instructions to Chamberlain were clear: "Hold at all hazard." Chamberlain had the high and dangerous honor of anchoring the left flank of the entire *Army of the Potomac*. Within minutes his men were hotly engaged.

With the 47th Alabama, Colonel William C. Oates' 15th Alabama had been assigned to cover the right of Law's Brigade as it swarmed up Plum Run in an attempt to flank Ward's brigade in the Devil's Den. Advancing to Law's right,

Joshua L. Chamberlain

A native of Maine, Chamberlain (1828-1914) attended a military secondary school and then graduated from Bowdoin College and Bangor Theological Seminary. In 1855 he became professor of rhetoric and modern languages at Bowdoin. In 1862 he was granted a leave of absence to study abroad, but instead joined the *20th Maine* and was appointed lieutenant colonel. He participated in 24 engagements with his regiment, but it is for his heroic defense of Little Round Top on 2 July that he is chiefly remembered. The action, which eventually resulted in his being awarded the Medal of Honor, remains an object of study in leadership classes in the United States Army today. Appointed a brigadier general by Grant at Petersburg, at the end of the war he was detailed to accept the surrender of the Army of Northern Virginia. Brevetted a major general of volunteers, he refused a commission in

the Regular Army and returned to his native state, where he was elected governor. He served three one-year terms (1867-1870) and then returned to Bowdoin as professor of mental and moral philosophy. He was soon appointed president of the college, serving for 13 years, and thereafter continued to lecture until 1885, allegedly having taught in virtually every department in the college. Meanwhile, he also served in the state militia, rising to major general, made a considerable fortune in business, wrote a number of books, and held several government posts. At Chamberlain's home in Portland, which was so decorated with war souvenirs as to be something of a Civil War museum, there was a magnificent wooden dining table in which he invited distinguished veterans of both sides to carve their initials. Perhaps the ultimate citizen-soldier of the war.

the regiment had swept aside some Union skirmishers and begun to scale Big Round Top in the face of light opposition from elements of the *2nd U.S. Sharpshooters*. As Oates' men reached the crest of the craggy hill, the Union troops fell back, disappearing. Thinking this portended an attack by a superior Union force, Oates halted his men. Just then one of Law's staff officers rode up, demanding to know why the regiment had halted. Oates explained his misgivings. The staffer brushed them aside, insisting that Oates obey orders and continue the advance. Oates therefore got his weary men to their feet and led them forward once more. Minutes later, at about 1630 hours, the 15th Alabama had reached the foot of

Big Round Top, coming out into the narrow, lightly wooded valley which separated it from Little Round Top. From there, Oates was surprised to spot what appeared to be a major Union wagon park standing completely unguarded not 600 yards to the east. As he advanced on it, he came under heavy fire from the base of Little Round Top. Thinking the Federal skirmishers had retired thence, he ordered his regiment, approximately 500 men, to sweep them away. As they advanced, the Alabamians ran into Joshua Chamberlain and some 385 determined men of the *20th Maine*.

While the 350 men of the 47th Alabama to his left engaged the *83rd Pennsylvania*, with about 300 men, and Chamberlain's right, Oates threw his regiment up the steep 45 degree, rocky, scrub, and tree covered slope of Little Round Top against Chamberlain's left, hoping to crush it and roll up the entire Union position. His men drove off Chamberlain's *Company B*, which was trying to form a skirmish line covering the regiment's left. As *Company B* and some of the sharpshooters rallied behind a stone wall at the eastern end of the little valley, the Confederates attacked Chamberlain's line. The fight grew hotter as Oates made repeated uphill assaults against the *20th Maine*, each time being beaten off into the thick woods. Despite this, Confederate pressure did not let up. Hand-to-hand fighting developed, but the Maine Staters held firm. Great heroism and dedication were displayed on both sides. Chamberlain kept his men well in hand, prepared to meet each new attack, and Oates was repeatedly able to rally his men and lead them forward. But finally, the limits of endurance were reached. Battered by both the *83rd Pennsylvania* and the *20th Maine*, the 47th Alabama fell back. Oates' men wavered, thirsty, and exhausted. Meanwhile, having fired over 20,000 rounds, the *20th Maine* was nearly out of ammunition. Realizing that another Confederate attack might succeed, Chamberlain ordered his men to charge with bayonets. Beginning from the left, the men rose from their places and stormed forward. Covering the 30 yards or so which separated them from the 15th Alabama in less than a minute, they threw the enemy into confusion, who fell back hotly pursued

An engraving of a wartime sketch showing Brigadier General Gouverneur K. Warren observing the Confederate advance from the summit of Little Round Top. Although the general is rather nattily turned out, the physical details of the position are remarkably accurate.

Colonel Joshua L. Chamberlain, commander of the 20th Maine, held a remarkable defense of the extreme Union left at Little Round Top.

A wartime view of the lines on the evening of 2 July, looking north and westwards from the summit of Little Round Top.

by the determined Yankees. As they did, *Company B* and the sharpshooters returned to the fray, rising from behind their protective stone wall to deliver a shattering volley into the flank of the retreating Confederates. The 15th Alabama broke and, as Oates himself put it, "we ran like a herd of wild cattle." They were closely pursued by Chamberlain's victorious men, who bagged over 400 prisoners. From start to finish, the fight for Little Round Top had taken little more than an hour. In one of the hottest combats in American history, the *20th Maine* had suffered about 30 percent casualties while the 15th and the 47th Alabama had been shattered, losing upwards of 40 percent of their men. But even as Chamberlain's men swept the enemy from the southern face of Little Round Top, the defenses on the western side of the hill began to crumble.

West of Little Round Top, Brigadier General John Ward's brigade was putting up a stiff resistance to Law's and Robert-

Looking across the valley of de from little Round-Top.

An engraving of a postwar photograph taken from the summit of Little Round Top, looking westwards and northwards, in an arc from the Peach Orchard to Cemetery Hill. Note the monument, to the 91st Pennsylvania, one of the first of the scores which now dot the field.

son's Confederate troops in the Devil's Den. As Colonel William F. Perry of the 44th Alabama noted, this was an area in which "Large rocks, six to fifteen feet high, are thrown together in confusion over a considerable area, and yet so disposed as to leave everywhere among them winding passages...." There were hundreds of places which could shelter a soldier, scores of routes among the boulders, and lots of timber and rock with which to erect breastworks, all features which favored the defense. In an effort to get past Ward, Law's men had been slipping around his left up Plum Run. Some of them had begun climbing the western face of Little Round Top. This entangled them in the fighting there. Gradually the rest of the brigade was drawn into the struggle which grew even more intense. Observing from the Devil's Den, Colonel Perry remarked that Little Round Top "resembled a volcano in eruption"; Captain Praxiteles Shaw, a preacher turned soldier with the 5th Texas, put it differently, but with equal poetry when he said that Yankee "fire came down the hill in blizzards." Holding the western side of the hill was the *16th Michigan*. A seasoned veteran regiment with good officers, it was the weakest outfit in Vincent's brigade and it was in the wrong place. With perhaps 250 rifles in the line, it could not cover Vincent's flank, which hung some 450 yards to the left of Ward's brigade, above Plum Run and the wooded ground at the foot of Little Round Top. Struck by elements of two brigades, Law's and Robertson's, which advanced up the bare steep western slopes with great courage and elan, the *16th Michigan* recoiled, some say broke and ran, with its commanding officer Colonel Norval E. Welch in the lead. The *44th New York*, to the left of the Michiganders, tried to refuse its flank, bending it back and fighting on two fronts. With his right in imminent danger of collapse, Vincent came up exhorting and rallying his men, only to fall mortally wounded. Little Round Top hung ripe for the plucking as Confederate forces advanced up its western face. But Gouverneur Warren was on the scene once more, having returned to the hill shortly after Vincent's men had arrived. Even as Vincent fell, on the craggy summit Warren had been supervising the emplacement of the 10-pounder Parrott guns from Lieutenant Charles E. Hazlett's *D Battery* of the *5th Artillery*, a Regular Army outfit. As Hazlett's guns began to open up on the Confederate troops beneath Little Round Top and assailing Devil's Den, word reached Warren that Vincent had fallen and that the entire position was crumbling. From the crest of Little Round Top, Warren spotted a brigade on the march towards the Peach Orchard. He rode down to it with his aides and was pleased to discover that it was his old brigade from the *2nd Division* of *V Corps* now under Brigadier General Stephen H. Weed. Sykes had ordered Weed to reinforce Little Round Top. However, while on the march the latter had been

Lieutenant Charles E. Hazlett's D Battery, 5th Artillery *in action on the summit of Little Round Top.*

intercepted by Sickles, who had peremptorily instructed Weed to march on the Peach Orchard, where he was under considerable pressure. Weed had ridden ahead and the brigade was temporarily under Colonel Patrick H. O'Rorke of the *140th New York*. Warren countermanded Sickles orders and within minutes the brigade began streaming up the north side of Little Round Top. At that moment Warren's brother Edgar, one of Weed's aides, came up demanding to know what was going on, as the brigade was needed elsewhere. Warren explained the situation and the brothers agreed that O'Rorke's regiment should remain with Warren, while the balance of the brigade resumed its march for the Peach Orchard. O'Rorke, who had recently graduated from West Point at the head of his class, quickly ran his 450 men up the steep slope to the summit of Little Round Top, where Hazlett now had six guns in position. Without pausing—and with

Brigadier General Evander McIvor Law, a brigade leader in Hood's Division.

empty rifles—the regiment plunged down the other side with fixed bayonets. Within minutes Weed came up with the balance of his brigade, Sykes having overruled Sickles. The attack by nearly 1,500 men threw the Rebels back in bloody disorder. O'Rorke fell at the head of his troops, Hazlett as he commanded his battery, Weed as he rushed forward. There were a few minutes more of heavy fighting and then it was over; the Rebels gave up their efforts to take the hill. Soon after, the balance of Sykes' *2nd Division* came up, firmly anchoring the Union left. Little Round Top was secure. But the situation along Sickles' front was still fluid, and the Union could still lose the day.

During the fighting for Little Round Top, Sickles' *III Corps* had been repeatedly battered back in a bloody, confused series of actions characterized by an erratically drawn front, a continuing commitment of reinforcements, a constant juggling of regiments from place to place in the line, and a near total absence of higher direction. Sickles stripped Colonel George C. Burling's brigade from Major General A.A. Humphreys' *2nd Division* on his right to reinforce Major General David B. Birney's *1st Division* on his more exposed left, bringing the latter up to nearly 6,500 men. Birney, in turn, did much the same thing, stripping regiments from Colonel

Evander McIvor Law

A native of South Carolina, Law (1836-1920) attended the Citadel, where he served as an instructor in literature before graduating in 1856. He became a teacher in secondary schools and on the outbreak of the war recruited a company for state service from among his students at the Tuskegee Military High School. The company served at Pensacola and was then incorporated into the 4th Alabama, the men of which elected Law lieutenant colonel. Severely wounded at Bull Run, upon his recovery Law was made colonel and led the regiment in the Peninsula, during which he also temporarily commanded a brigade, and in the Second Bull Run and Antietam campaigns. He was promoted brigadier general in October of 1862 and given a brigade in Hood's Division at Fredericksburg and in the Carolinas. At Gettysburg, it was Law's brigade which attempted to storm Little Round Top and the Devil's Den on the afternoon of 2 July. During this action Law took command of the division on the wounding of John B. Hood and led it for the balance of the campaign. Law went West with the division later in 1863 and again assumed command when Hood was wounded at Chickamauga. Returning to Virginia, he led his brigade with considerable skill and courage during the opening phases of the Grant's Overland Campaign of 1864 until wounded at Cold Harbor. Law then served in administrative posts in the Carolinas for a time and in the closing months of the war commanded a cavalry brigade in General Joseph E. Johnston's Army of Tennessee. After the war Law settled in Florida, where he was a newspaperman, a vocal advocate of educational reform, and was active in Confederate veterans organizations.

Philippe Regis de Trobriand's brigade and Burling's, to reinforce Brigadier General Charles K. Graham's 1,500 men covering the Peach Orchard on his right and Brigadier General John Ward's 2,200 on his left. Sickles even secured Meade's permission to draw troops for elements of *II Corps* and *V Corps*, though Sykes soon reasserted his authority in the latter case. This practice did little for the morale or the cohesion of the affected troops. Forced to participate in ill-coordinated actions under unknown officers with strange units on their flanks, often on broken wooded ground, the men could not perform to the best of their abilities. Able officers, but with few troops, both Burling and de Trobriand were unemployed for much of the fight, when they might have usefully commanded sections of the front. The result of all this was that the defense suffered. Another result was that it became extraordinarily difficult to elucidate the movements of the units involved, as they struggled with each other, or even to get a clear notion of the sequence of events in the front of *III Corps* on the afternoon of 2 July.

Ward's brigade to the west of Little Round Top had been under heavy pressure from the moment the Rebel guns opened up on the Peach Orchard in the center of *III Corps*. But this had begun to slacken during the heaviest part of the fighting on Little Round Top, for most of Law's and Robertson's troops became involved there. As a result, Robertson found himself facing Ward with but two regiments. He requested reinforcements from Hood, only to discover that the latter was down with a shattered arm and had not yet been replaced, for the next ranking officer in the division was Law, who was at Little Round Top. Robertson therefore requested generals Henry Benning and George T. Anderson, commanding Hood's other brigades, to reinforce him with their troops, while sending messages to Longstreet. Benning's brigade was immediately available, for he had become lost and followed Robertson rather than Law. G.T. Anderson brought his men up as well. With two full brigades and a portion of a third, Robertson renewed the attack on Ward's brigade, and took on de Trobriand's weakened one to Ward's right as well. Apprised of the renewed Confederate onslaught against Ward, Sickles committed three regiments from Burley's brigade. But even these could not stem the Rebel tide. Despite favorable terrain, Ward and de Trobriand finally began to fall back, giving up Devil's Den and the southern end of the Wheatfield. To make the situation even worse, Sickles was wounded soon after ordering Burley's men forward, leaving the corps to fight on without any overall direction. As this occurred, reinforcements arrived to bolster the Federal line in the form of two brigades from Caldwell's division of *II Corps* and two more from Ayres' division of *V Corps*. These launched a counterattack, which halted and then partially threw back the enemy. By about 1730 the line had temporarily stabilized. Then two brigades of Brigadier General James Barnes' *1st Division* of *V Corps* began to come up, swinging around the Union right, and endangering the left flank of G.T. Anderson's attacking Confederates. Hood's Division, perhaps 6,800 strong at this point, was facing over 10,000 men and imminent disaster.

The fierce fighting on his right had begun to concern Longstreet. He ordered up McLaws' Division, which had dawdled in commencing its own attack scheduled for 1700. Brigadier General Joseph B. Kershaw's South Carolinian brigade went in first, some 2,200 men, at just about 1745, followed almost immediately by Brigadier General Paul J. Semmes' 1,300 Georgians. The two brigades drove towards the 400-yard gap between de Trobriand's right and the Peach Orchard held by Graham's Pennsylvanians. It was into this area that Colonel William S. Tilton and Colonel Jacob B. Sweitzer, of Barnes' *1st Division*, had intruded their brigades, thereby threatening to flank G.T. Anderson with 2,000 men. Finding himself now flanked in turn by Kershaw, Tilton pulled his small brigade back, twisting his right around until it faced westwards. This exposed Sweitzer's flank. Barnes ordered Sweitzer to break off his attack and fall back, which the latter did despite an inclination to fight it out. Kershaw advanced under artillery fire into the ground vacated by Tilton and Sweitzer, only to have his right flank smashed by the brigades of Brigadier General Samuel K. Zook and Colonel John R. Brooke, some 1,800 men from Caldwell's division of *II Corps*. With his right crumbling, Kershaw called Semmes to his aid; the latter's brigade came up quickly though Semmes fell mortally wounded as it did. Now under Colonel Goode Bryan, Semmes' men advanced under heavy fire until halted. At this moment, about 1800, perhaps 15-20 minutes after Kershaw and Semmes had advanced, Brigadier General William Barksdale's brigade of 1,600 Mississippians leaped forward, smashing into Brigadier General Charles K. Graham's brigade at the Peach Orchard. The 1,400 Georgians of Brigadier General William T. Wofford's brigade followed

Charles K. Graham

Born in New York City, Graham (1824-1889) joined the Navy as a midshipman at age 17. He served in the Gulf Squadron during the War with Mexico, and resigned the following year to return to his native city, where he studied both law and engineering, qualifying to practice both. He worked with architect Frederick Law Olmstead in laying out Central Park. The outbreak of the war found him supervising the construction of the dry docks at the Brooklyn Navy Yard, and he immediately enlisted in Dan Sickles' *Excelsior Brigade*, along with 400 other dockyard workers. Graham soon became colonel of one of the brigade's regiments, *74th New York*, serving in the Peninsula campaign. He spent the next year largely engaged in recruiting duty due to ill-health, but returned to command a brigade in Major General David G. Birney's division of Sickles' *III Corps* as a brigadier general of volunteers in time for Chancellorsville. At Gettysburg his brigade was severely battered holding the Peach Orchard and he himself was wounded and captured. Exchanged in September, he was later sent to command the river gunboats in the *Army of the James*. At the end of the war he was brevetted major general of volunteers and returned to civilian life. Graham had a distinguished career as a civil engineer in New York City until his death.

within minutes with Longstreet himself in the lead. As the troops cheered, Longstreet plunged forward, waving his hat, and crying, "Cheer less, men, and fight more!" Barksdale's men swung to their left and proceeded to take A. A. Humphreys' division in the flank along the Emmitsburg Road, while Wofford's troops pressed eastwards and then to their right. Graham's men were driven from the Peach Orchard in considerable disorder, some 250 of them falling prisoners, including Graham himself, as Wofford's men linked up with Kershaw's forming a single battle line; some of the captives, temporarily stashed in the Sherfy house, would later be killed by Union artillery fire. Major James Dearing, of Longstreet's corps artillery, quickly got eight cannon forward and began to belabor the retreating Union troops. The situation on the Union front was now critical, for Hood's men had turned the left of *III Corps* at the same time that McLaws' had broken its center. Seven brigades—the equivalent of a full army corps—

124

James Dearing

Born on a plantation in Virginia, Dearing (1840-1865) was a member of the West Point class of 1862 when he resigned in April of 1861. He entered Confederate service as a lieutenant in Louisiana's Washington Artillery Battalion. He later commanded a Virginia battery in Pickett's Division during the Peninsula campaign and as a captain at Fredericksburg and Chancellorsville. Promoted major, he was given command of Pickett's artillery battalion, which he led at Gettysburg, participating in the grand battery which bombarded Cemetery Ridge on 3 July. He then commanded cavalry under Pickett in North Carolina as a temporary colonel during the winter of 1863-1864. In April of 1864 Dearing was made lieutenant colonel and given command of the horse artillery of the Army of Northern Virginia. He was soon after promoted brigadier general and given a brigade of cavalry on the North Carolina coast. He fought with great distinction during the Petersburg campaign in 1864 and 1865, serving until the evacuation of Richmond. On 6 April 1865, during the retreat to Appomattox, Dearing engaged in an exchange of pistol shots with a Union general, slaying his opponent but falling mortally wounded. He lingered on in a hospital, where he was paroled by a former West Point classmate, Brigadier General Ranald Mackenzie, and succumbed on 23 April, the last Confederate general to die in the war.

---

were in danger of being enveloped on the left of *III Corps*, and Humphreys' division, already engaged frontally with Major General Richard Anderson's Confederate division, now lay exposed to a flank attack from its left.

Brigadier General John C. Caldwell tried to secure the Union line running from the Wheatfield to Devil's Den. He led Brooke's and Sweitzer's brigades in a charge across the corpse-filled field, supported on his left by two brigades of Regulars from Brigadier General Romeyn B. Ayres' *2nd Division, V Corps*, who attacked Devil's Den. The troops fought hard, driving back the enemy in a bloody charge. Then, even as the victorious Federal troops pressed the attack, they were struck on their right flank by Wofford's Rebels emerging from the Peach Orchard. The attack faltered, as bloody hand-to-hand fighting developed. Pressured on both front and flank, the Union troops suffered heavy casual-

Having snatched up the colors of the 1st Pennsylvania Reserves, *Union Brigadier General Samuel W. Crawford leads his division into the gap between the Devil's Den area and Little Round Top. Not shown in this wartime illustration is that the regimental color bearer Corporal Bertless Slott, unwilling to abandon his duty, grabbed the leg of the general's trousers and hung on, racing beside him as he charged the enemy.*

ties and gave way. Some units broke, fleeing before the advancing enemy to the shelter of Cemetery Ridge and Little Round Top, hundreds of yards to the rear. The front line had dissolved. Three full divisions had been badly battered. The entire Union left was now at risk. Ordering elements of *VI Corps* and *XII Corps* to help fill the gap in the Union lines south of *II Corps*, Meade placed Hancock, of *II Corps*, in command of the entire flank. Hancock thrust Brigadier General Samuel W. Crawford's *V Corps* division of the *Pennsylva-*

nia Reserves into the gap to cover the Devil's Den area and Little Round Top from the west.

When Crawford reached down from his mount to grab the colors of the *1st Pennsylvania Reserves* in order to lead it forward, the color bearer Corporal Bertless Slott, unwilling to abandon his duty, grabbed the leg of the general's trousers, and hung on, racing beside him as he charged the enemy. Crawford soon had his troops in position with one brigade on the hill, and the other massed to its north. As thousands of fugitives streamed past, the Pennsylvanians stood patiently awaiting the enemy for 20 minutes. When the tight Rebel columns came into view out of some woods in the direction of the Wheatfield, Crawford ordered an attack. Thousands of muskets spoke as the Pennsylvanians discharged two volleys, then, with Crawford in the lead, charged, racing downhill, and sweeping the advancing Confederates back once more to the edge of the Wheatfield, halting the threat to the Union left. If the center and right of the *III Corps* front could be held, disaster might yet be averted.

Over on *III Corps'* right, A. A. Humphreys had been doing well most of the afternoon, being only lightly engaged against elements of Confederate Major General Richard Anderson's division to his front. As the crisis had developed on the Union left, he had been subject to conflicting orders. When Sickles was wounded, Major General David B. Birney, commanding the corps' *1st Division*, had assumed command. Rather than direct the fighting on the left flank of the corps, Birney had abandoned his division and gone off to meddle with Humphreys' dispositions. The aggressive Humphreys had planned to counterattack into the face of the impending enemy attack, catching them off balance. Birney canceled this, ordering Humphreys instead to fall back somewhat, and draw back his left in a maneuver which Birney fondly hoped would restore the Union line as soon as his own division could similarly refuse its right. But even as Humphreys ably shifted his division to form this new line—despite the fact that there was no one to his right on whom he could form up—Graham's brigade collapsed in the Peach Orchard. This

Andrew A. Humphreys

Scion of the famed Humphreys of Philadelphia, the distinguished naval architects who designed the USS *Constitution* and many other ships of the Old Navy, Andrew (1810-1883) graduated from West Point 13th in the class of 1831. Most of his military career was spent in the Corps of Topographical Engineers, conducting hydrographic surveys on the Mississippi River. At the outbreak of the war he was appointed an aide to Major General George B. McClellan. Promoted brigadier general of volunteers, he served as chief of Topographical Engineers of the *Army of the Potomac* during the Peninsula campaign. In the autumn of 1862 he was given a division in the *V Corps*, leading it with distinction at Antietam, Fredericksburg, and Chancellorsville. Given a division in the *III Corps* in time for Gettysburg, Humphreys soon proved himself a soldier of uncommon ability. By strong leadership and great courage, he safely extracted his division from the collapse of the corps which resulted from Major General Daniel Sickles' poor deployment on 2 July.

After Gettysburg Humphreys was made major general of volunteers and brigadier general by brevet in the Regular Army, and appointed chief of staff to Meade. He served in this post with great skill until late in 1864, when he was given Major General Winfield Scott Hancock's *II Corps*. He led this corps through the closing campaigns of the war, receiving the brevet of major general in the Regular Army. After the war he was promoted to brigadier general in the Regular Army and named chief of engineers, which post he occupied until his retirement in 1879. In retirement Humphreys wrote *The Virginia Campaign of 1864 and 1865*, still a useful work. A very fine officer, skilled both in combat and in administration, Humphreys is believed to have participated in 70 engagements during his military career.

---

put Barksdale's Mississippians on his exposed left flank, even as Anderson threw two brigades against his front and right. Birney now intervened once more, ordering Humphreys to fall back all the way to Cemetery Ridge over 850 yards in his rear. This was an unwise maneuver, for the division had to retire while receiving the full attentions of three Confederate infantry brigades and a considerable number of cannon. Humphreys made the best of it. He pulled his men back slowly, repeatedly halting them so that they could get off a few volleys. Despite heavy casualties, the troops held to-

David B. Birney

The son of the prominent Abolitionist James G. Birney, James B. Birney (1825-1864) was born in Alabama but the family subsequently relocated to Ohio. Graduating from Andover, he entered business and studied law, practicing in Philadelphia from 1856. Birney read widely in military subjects and when the war began was appointed lieutenant colonel of the *23rd Pennsylvania*. When the regiment reenlisted for three years he was named colonel and soon after appointed brigadier general of volunteers. He commanded a brigade in the *III Corps* during the Peninsula campaign, after which he was court martialed but acquitted on a charge of disobedience of orders and restored to duty in time for the Second Bull Run campaign. At Chantilly he succeeded to the command of Major General Philip Kearny's division when the latter was killed. Promoted major general for distinguished conduct at Chancellorsville, Birney performed poorly in an admittedly bad situation at Gettysburg, where he assumed command of the *III Corps* upon the wounding of Dan Sickles. In Grant's Overland Campaign of 1864 he performed well, and was to have had the *X Corps* but took sick with malaria and died. Birney was a good, if uninspired commander. At Gettysburg he was thrust into an almost impossible situation which he made worse by doing the proper thing, going off to assume command, rather than the practical one, remaining with the troops. His older brother, William Birney, rose to a brevet major generalship during the war.

gether well, inspired by Humphreys' heroic example as he rode among them, exhorting them to greater efforts, and issuing orders. Artillery from Cemetery Ridge began to support him, and several regiments came out to lend a hand. The entire maneuver took perhaps 20 minutes, and the division suffered 1,200 dead and wounded, over 35 percent casualties, but it came out intact with morale sound and still full of fight. Even as Humphreys' men reached the security of Cemetery Ridge, other elements of *III Corps* managed to escape from the disaster as well.

About 50 pieces of artillery had been concentrated in the area just east of the Peach Orchard and north of the Wheatfield. Having supported the fight all afternoon, these batteries were suddenly exposed to capture with the collapse of Graham's brigade at the Peach Orchard and the disintegra-

William Barksdale

Born in Tennessee, Barksdale (1821-1863) attended the University of Nashville. Moving to Mississippi, he studied law and then edited a pro-slavery newspaper in Columbus. He served as a volunteer in the Mexican War, rising to lieutenant. In 1851 he was elected to the House of Representatives and served until his resignation in January of 1861 upon the secession of Mississippi. His fame as a champion of the righteousness of slavery and states' rights for a time made him a rival to Jefferson Davis for the presidency of the Confederacy, but upon the latter's election Barksdale became state quartermaster general and then entered Confederate service as colonel of the 13th Mississippi. He led his regiment with considerable ability during the Bull Run and Peninsula campaigns, during which he was given a brigade of Mississippians in McLaws' Division of Longstreet's First Corps. He led this with distinction at Antietam, Fredericksburg, and Chancellorsville, where they were instrumental in holding back greatly superior forces late in the day. At Gettysburg on 2 July, Barksdale led his brigade in the attack against part of Sickles' *III Corps* in the Peach Orchard, where he was mortally wounded and captured.

---

tion of the Union line. Coordinated by Lieutenant Colonel Freeman McGilvery they held off the advancing Rebel tide as long as possible, offering an island of resistance in a sea of shattered divisions and regiments. At about 1800 they fell back some 250 yards to the vicinity of Trostle's farm. There they engaged in a number of desperate rear-guard actions against Kershaw's, Barksdale's, and Wofford's Confederates. As pressure grew, McGilvery fell back once more. Under cover of the half-dozen 12-pounder Napoleons of Captain John Bigelow's *9th Massachusetts Battery*, he formed a new gun line on Cemetery Ridge about 400 yards to the rear. Bigelow's men fought on without infantry support until they were down to two guns. Then, with most of their horses dead and nearly all of their canister exhausted, and with Rebel infantry closing in on their flanks and front, they pulled back to McGilvery's new line. As the enemy closed on his new artillery line, elements of Kershaw's Brigade overran several batteries. The *141st Pennsylvania* came up and retook them in a bayonet attack. In the end, only four guns were taken by the

Brigadier General William Barksdale was mortally wounded while leading his troops on the attack at Gettysburg. He was captured and died the next day in captivity.

enemy, though many more were disabled or destroyed. The balance were brought safely away and by about 1915 were lined up along Cemetery Ridge, just as Hancock came up to their support, personally leading a brigade from his *3rd Division*. Hancock threw three regiments into a frontal counterattack downhill against Barksdale's troops, while a fourth went in to recover a battery which the 21st Mississippi had overrun on the left. In the bloody clash which followed Barksdale's Mississippians were thrown back and he himself was killed by the desperate charge of the *111th, 125th*, and *126th New York Regiments*, while the *39th New York* (a polyglot outfit of Germans, Italians, Poles, and Hungarians) retook the guns, and three Rebel flags in the bargain. The Federal line began to solidify on Cemetery Ridge as fragments of regiments drifted back. Two brigades which Meade had ordered up from *XII Corps* began to arrive, led personally by acting corps commander Brigadier General Alpheus Williams. Williams threw his two leading regiments into the fighting, driving off still more Confederate troops, and recovering several more guns. Then he pressed hard against the Rebels, pushing forward in the gathering dusk almost to the Peach

A Rabbit's Moment in History

A rabbit played a minor role in Gettysburg on the second day. As the 3rd South Carolina passed among the buildings of the Rose farm that hot afternoon a terrified rabbit fled before it. Seeing the animal, one of the troops called out, "Go it, old fellow; and I would be glad to go with you, if I hadn't a reputation to sustain!" Unfortunately fleeing *before* the troops meant heading the same way they were going, towards the enemy. Within minutes, the terrified rabbit ran right into the men of the *118th Pennsylvania* as they were lying in wait for the Rebels in some woods on the southeastern edge of the Wheatfield. As the South Carolinians behind were making more noise than the Pennsylvanians before, the bunny plunged on, momentarily landing on the neck of one of the troops before bounding away. As the rabbit fled to the rear, its moment in history at an end, the soldier jumped up, crying, "Oh! I'm shot! I'm a dead man!" much to the amusement of his comrades, at least until the Rebels were upon them.

Orchard nearly 850 yards beyond Cemetery Ridge, while he deployed the rest of his troops along the ridge itself. Thus the final gap in the Union left was plugged, and a continuous front established from Little Round Top to Cemetery Hill.

Meanwhile the fighting became more general. Even as Humphreys withdrew his division under pressure to Ceme-

Union artillerymen in action during the desperate rearguard fighting which occurred before Seminary Ridge as evening fell on 2 July.

Engraving made from photographs taken a day or two after the battle, showing Trostle's farm, the post of Captain John Bigelow's **9th Massachusetts Battery,** *which covered the Union retreat to Cemetery Ridge on the evening of 2 July, losing in the process most of its men, all of its horses, and all but one gun.*

tery Ridge, A.P. Hill launched his Third Corps against it and the northern end of the Cemetery Ridge line. Soon the entire front of *I Corps* and *II Corps* was hotly engaged. Small units performed prodigiously, making limited attacks at every opportunity. The fighting grew intense as Hill fed brigade after brigade into the attack. Wilcox's 1,700-man brigade of Anderson's Division was halted by a desperate charge of 262 men of the *1st Minnesota*, which sustained one of the highest regimental losses in a single action in American history, fully 215 men—82 percent—became casualties, including Colonel William Colville, Jr., who had been released from arrest to participate. Meade pulled the *2nd* and *3rd Divisions* of *I Corps* out of reserve and plugged them into the lines, personally leading their skirmishers forward. A series of local counterat-

Spangler's Spring, at which a purported "truce" took place on the night of 2-3 July, from an engraving made from a wartime sketch.

tacks resulted, gradually merging into the semblance of a general counterattack. The Confederate attack faltered. As darkness began to fall, Hill's brigades fell back, closely pressed by Hancock's regiments. And as the Union forces threw the Confederates back to Seminary Ridge, heavy fighting began to develop on their right as well, in front of Cemetery Hill and Culp's Hill.

Save for some desultory shelling of Union positions on Cemetery Hill and Culp's Hill, Richard Ewell's Second Corps had been virtually unengaged for most of 2 July. Although Lee had instructed him to attack if he thought conditions suitable, Ewell appears to have had no clear notion of what he was going to do. At about 1600 hours, 16 pieces of artillery began to lay down a more vigorous barrage against the eastern side of Cemetery Hill in response to the opening rounds of Longstreet's offensive on the Confederate right. As Federal artillery began counter-battery fire against the Rebel guns, Yankee infantrymen posted in the vicinity prepared to meet a general assault. But none came. Instead, the artillery duel dragged on for hours, with the more numerous and heavier Union guns gradually gaining the upper hand. At about 1800 hours, as the gunners began to slacken off their fire, Ewell briefly toyed with the idea of launching an attack. Having second thoughts, he soon gave it up. This permitted Meade and Slocum to strip much of *XII Corps* away to reinforce the shattered Union left. Some time later, Ewell again decided to attack. He issued the appropriate instructions to his subordinates, but thereafter appears to have done little to supervise the operation. As a result, Major General Robert E. Rodes' division on Ewell's right completely failed to get into action. Major General Edward Johnson's division, over on Ewell's left, did better, and Major General Jubal Early's, in Ewell's center, did the best of the three. The attack began at about 2000 when 32 guns opened up on the Union positions. Johnson attacked Culp's Hill from the northeast and east. He went in with three brigades, being forced to use the famed Stonewall Brigade to cover his left where Union cavalrymen under Brigadier General David Gregg were constantly probing his rear. The attack did not go well. The troops encountered great difficulties in getting across deeply cut Rock Creek under fire. By the time they finally cleared the creek, pushed back the Yankee skirmishers, and began to grapple with the main line of resistance, it was already dark, sunset having been at about 2030. The Union troops, the battered veterans of Wadsworth's division of *I Corps*, with a brigade from *XII Corps* to its right, were posted atop steep slopes and rocky cliffs which they had strengthened by breastworks of logs, rocks, and earth. Johnson's principal effort hit Brigadier General George S. Greene's brigade of the *3rd Division* of *XII Corps*, numbering little more than 1,400 all told, with a long front to cover. Although already under attack himself, Wadsworth immediately reinforced Greene with two weak regiments, which were soon followed by several more from Carl Schurz' *XI Corps* division to the left of Culp's Hill, and one from *II Corps*, which returned almost as soon as it arrived. These were few in number, bringing Greene's total up to at best 2,200 men, but they were enough, given his excellent position. By careful supervision of his troops and by rotating regiments into reserve from time to time to rest and resupply, Greene managed to hold the attacks of Brigadier General John M. Jones' Virginians and Colonel J. M. Williams' "Louisiana Tigers" against his front. Over on Greene's right, however, Confederate Brigadier General

George H. Steuart's brigade achieved considerable success, for it attacked against an unoccupied section of the defenses which had been evacuated earlier in order to free troops to go to the support of Sickles' beleaguered corps. Greene refused his right and moved troops to cover it. But the real defense was provided by the lunar landscape of the southern end of Culp's Hill and the ground falling away from it, which was full of rocky outcroppings, woods, depressions, and hollows. The Confederates advanced cautiously in the dark, fearing a trap. And as they did, Brigadier General Thomas Ruger marched up with his division, which had been taken from *XII Corps* to bolster the left, but was now returning. Ruger's division deployed quickly, covering the Rebel positions from across Spangler's Spring, a small brook running down from Culp's Hill into a broad swampy area up against Rock Creek. The lead elements of the other division of *XII Corps*, that of Brigadier General John W. Geary, began to come up at about 2300 hours after having gotten lost and wandered aimlessly in the Federal rear for several hours. By midnight, Geary's men had deployed to cover Steuart's front. Neither Union commander thought it wise to attempt an attack against Steuart in the dark, particularly in as much as his men were occupying earthworks which their own men had built, but then been forced to abandon in order to march off to bolster the line on Cemetery Ridge.

In Ewell's center was his best division, Major General Jubal Early's. Early attacked Cemetery Hill at around 2000 hours, with about 3,500 men on a two-brigade front. As they emerged from cover the troops came under heavy artillery fire from four batteries. Veterans all, Early's men kept coming, crashing into a line of Union infantrymen in the hollow ground at the foot of the hill. The defenders, from *XI Corps*, poured in a deadly fire. But *XI Corps* had been heavily engaged on the previous day and was weak and its men were tired. Colonel Leopold von Gilsa's brigade of Brigadier General Adelbert Ames' *1st Division*, on the corps' right, was down to little more than 650 men, and Colonel Andrew L. Harris' brigade, at right angles on its left, was even weaker.

Major General Jubal Early. On 2 July he urged corps commander Ewell not to attack Culp's Hill. After the war, he attempted to smear comrade Longstreet by blaming him for the defeat at Gettysburg.

As Brigadier General Harry T. Hays' Louisianans and Colonel Isaac Avery's North Carolinians slammed into them, the two brigades crumbled away, officers and men alike fleeing to the rear despite efforts to rally them by some steadfast artillerymen. As his infantry got onto the top of the hill, Early began to move some artillery forward. The Confederate pieces opened up. A heated hand-to-hand fight developed in the cemetery as Union gunners tried to beat off Confederate infantrymen. Placing his hand on a piece, one triumphant Rebel is supposed to have yelled, "This battery's ours!" only to receive the guttural reply, "*Nein*, dis battery is *unseres!*" A crisis was at hand. Major General O.O. Howard commanding the corps was conferring with Major General Carl Schurz, of his *3rd Division*, when the Rebels erupted onto the top of Cemetery Hill. Schurz gathered up two regiments immediately at hand, threw them into the fighting with fixed bayonets under Colonel Wladimir Krzyzanowski, and then personally followed them into the fray with his staff. Brigadier General Adolph von Steinwehr, his sector further to the left relatively secure, sent a brigade from his *2nd Division*.

Jubal A. Early

Born in Virginia, Early (1816-1894) graduated 18th in the West Point class of 1837, entering the artillery. After a tour of duty in Florida against the Seminoles, he resigned from the Army to practice law and engage in politics in his native state with considerable success. In 1861 he was a delegate to the Virginia secession convention, at which he voted against secession. Nevertheless, he became an ardent Confederate, entering the army as colonel of the 24th Virginia. He commanded a brigade under Beauregard during the Bull Run campaign and was promoted brigadier general. Continuing with his brigade he led it during the opening phases of the Peninsula campaign in 1862, until wounded. During the Second Bull Run campaign he returned to duty, commanding a brigade and during the Antietam campaign took over a division, which he led at Fredericksburg. Promoted major general in January of 1863, he held active commands until almost the end of the war. At Gettysburg on 2 July, he was among the officers who convinced Ewell not to attack Culp's Hill simultaneously with Longstreet's attack in the south, a matter which seriously affected the outcome of the day's fighting. During the battle of Spotsylvania, he assumed command of Richard Ewell's Second Corps. Promoted lieutenant general, he commanded the corps at Cold Harbor and then in July led it in a raid down the Shenandoah Valley and into Maryland with the intention of putting pressure on Washington, thereby causing troops to be pulled away from Richmond. He was partially successful and repeated the feat later the same month. The operation enjoyed less success as a strategic diversion, but the Confederates managed to plunder and burn along a broad front, spreading devastation as far as Chambersburg in Pennsylvania. Early was later trounced by Major General Philip Sheridan at Cedar Creek and his command was dispersed by Brigadier General G.A. Custer early in 1865. He was relieved by Lee and spent the remaining weeks of the war inactively.

After the war Early fled to Mexico, thinking to take service with the forces of the Emperor Maximilian, but soon went to Canada and then returned to Virginia, where he resumed his law practice and became involved in the corrupt Louisiana Lottery. He became the first president of the Southern History Society, wrote a number of books on the war, and engaged in a rather successful campaign to place all blame for the Confederate reverse at Gettysburg on Longstreet. He remained an ardent Rebel until his death. Nicknamed "Old Jube" and "Jubilee," Early was a fairly good commander, but independent minded and dishonest.

Attack of the Confederate Brigadier General Harry T. Hays' Louisiana Brigade on the Union XI Corps on Cemetery Hill. The Louisianans managed to gain a foothold on the hill, but were forced to retreat.

Colonel Charles R. Coster's New Yorkers and Pennsylvanians came up swiftly, delivered a couple of volleys into the enemy and threw them back on their right. At almost the same time, in response to a message from Howard, Colonel Samuel S. Carroll's brigade of *II Corps* came up. Carroll's men, from Indiana, Ohio, and West Virginia, decided the issue, as they threw the last of the Confederate troops off Cemetery Hill. Then they launched themselves over the edge and attacked downhill, clearing the enemy away from the foot of the hill.

Carl Schurz

A native of Rhenish Prussia, Schurz (1829-1906) was educated at the universities in Cologne and Bonn. In 1848 he served the Liberal revolution there as a junior officer. Upon the defeat of the revolution he fled, first to Switzerland and then to France. Expelled from France, he lived briefly in England and then migrated to the United States in 1852. He began a career as an orator and politician with a strong anti-slavery message, settling in Wisconsin. At the start of the war he was minister to Spain, but resigned in 1862 to urge immediate emancipation of the slaves. Lincoln demurred, but made Schurz a brigadier general of volunteers, a political move designed to enhance German-American support for the war. Schurz led a division with some ability in the Second Bull Run campaign. Made a major general in March of 1863, he and his division performed poorly as part of the *XI Corps* at Chancellorsville. At Gettys-burg he briefly commanded the corps, but neither it, nor his division performed with brilliance, though he did manage to hold the line on Cemetery Hill on 2 July. He accompanied his division West with the *XI* and *XII Corps*, where it again did poorly at Chattanooga. Relieved, Schurz spent the next year in administrative posts and stumping the country for Lincoln's reelection. Later he served ably as chief of staff to Major General Henry W. Slocum's *Army of Georgia*.

After the war Schurz became one of the most prominent political leaders in the country, championing a variety of progressive causes, and one of the few steadfast spokesmen for black civil rights in print, on the podium, and in the Senate. His political influence was enormous, particularly wherever there were large numbers of German-Americans. There are more memorials to his memory than to any other non-native-born American citizen.

Early's attack had failed. By 2230 hours it was all over, save for occasional rifle and artillery fire. Gradually the firing died down all along the line and the second day of the battle of Gettysburg was over.

It had been a good day for Confederate arms, but not the victory which Lee was seeking. Although great blows had been dealt the *Army of the Potomac* it had not collapsed, but had taken them and gone on to establish a substantial line of defense along Cemetery Ridge. Yet again the Confederate troops had fought splendidly, with but little to show for all their courage and devotion. They had not attained a decisive success. For the Union it had been a hard day, due primarily to Sickles' careless disposition of his corps. Nevertheless, the situation had been stabilized. Moreover, save for the collapse of Ames' division, the troops, by and large, had performed ably, and in many instances splendidly. Of course, as in the case of the Army of Northern Virginia, the *Army of the Potomac* had taken some heavy blows. About 13 brigades had suffered casualties of 30 percent or more, some of them having been so badly battered that they were unfit for further service. But the army had held together. The troops were still determined. That hot, humid night, as the armies bedded down in the moonlight, their commanders contemplated the morrow.

The Spangler's Spring Truce

On the eastern side of Culp's Hill, between the crest and Rock Creek, there is a deep, broad gorge in which there is a small spring called, after a local farmer, Spangler's Spring. A source of clean, cool water, the spring is partially responsible for the marshiness of Spangler's Meadow. There was much heavy fighting on Culp's Hill on the afternoon and evening of 2 July, particularly in the vicinity of Spangler's Spring. That night, according to many accounts, men of both armies sought water from the spring. As they crawled through the darkness, they often encountered, but pointedly ignored each other. A fine, touching tale, the story of the Spangler's Spring Truce often surfaces as an example of the basic humanity of men even when locked in the throes of battle. Unfortunately, it's completely untrue. The men who sneaked down to the spring to fetch water on that hot, humid night opened fire on each other repeatedly and enthusiastically, and skirmishing in the vicinity of the spring continued until dawn. The origins of the story about the truce are obscure, but the romantic tale appears to have begun to circulate about 20 years after the battle, when interest in it was rising, while the passions generated by the war were falling.

Who Won the Second Day?

The fighting on 2 July was once again a Confederate victory though an incomplete one at best, for the *Army of the Potomac* had escaped a potential complete disaster and retired to a secure defensive position. Again Lee's leadership had been wanting, nor had that of his subordinates been better. Longstreet's original plan of battle was probably technically superior to that which was eventually adopted, but he was somewhat careless in its execution, with the result that it was badly coordinated, and McLaws' men struck much later than ought to have been the case. Both Hill and Ewell should have been far more closely supervised. Hill had remained almost completely inactive for the entire day. Worse, by not getting Ewell to attack against the Union right at the same time that Longstreet struck their left, Lee permitted Meade to weaken himself on that side in order to restore his left flank after the collapse of the *III Corps*. Ewell's inexplicable lack of activity, bordering almost on the criminal, permitted a likely victory to turn into something approaching a drawn fight. Moreover, his attack on Culp's Hill later in the day, was foolish, unnecessary and hours too late to effect events in the center. Yet again the Confederate troops had fought splendidly, only to have little to show for all their courage and devotion.

Despite the reverse, Union leadership had done somewhat better. Meade had kept his head, appearing on all parts of the field to inspire the men and gain an understanding of the situation. He had confided in his most able subordinates, and eased the burden of command by placing Hancock and Slocum over the left and right wings of the army, respectively. They had done well, particularly Hancock, who seemed to be everywhere on the field at critical moments. Slocum handled the potential disaster on the Union right with considerable skill and Gouverneur Warren had been superbly effective, with his marvelous improvised defense of Little Round Top. Indeed, save for Sickles and Birney, the Union corps and division commanders had performed ably, and sometimes brilliantly. Sickles, of course, had been the cause of the disaster which involved his corps, and many other troops. Indeed, had the *III Corps* held its assigned position along Cemetery Ridge, it would have been ideally placed to take in the flank Longstreet's attack along the Emmitsburg Road. Birney contributed to the disaster when he left his division to go off and assume command of the disintegrating *III Corps* after Sickles had been wounded. Perhaps the most magnificent performance by a division commander had been that of Humphreys, who had kept his head and pulled his men out of the disaster in remarkably good order. Carl Schurz, commanding up on Cemetery Hill, also handled his division well. The lower ranking Union officers had generally, almost universally done well, most notably in the fighting around Little Round Top and the collapsing center. Their performance set the stage for the Union victory the next day.

Stuart's Ride—22 June-4 July 1863

As Lee moved northwards, he was legitimately concerned about the movements of the *Army of the Potomac*. Although his cavalry had done an effective job of denying the enemy information on his movements, it had been less successful in ascertaining the enemy's movements. By 22 June, when the Army of Northern Virginia had already begun to debouch from the Shenandoah Valley into central Pennsylvania, the *Army of the Potomac* had yet to cross to the north side of the river whose name it bore. Lee was generally uncertain as to the location and direction of Hooker's movements and decided to do something about it. He instructed his chief of cavalry, Major General James Ewell Brown Stuart, one of the finest troopers of the age, to observe the enemy closely. Should he detect any movement northwards on the part of the *Army of the Potomac*, Stuart was himself to advance northwards, placing his troops on the right of the head of the Army of Northern Virginia, Ewell's Second Corps, as it advanced, to cover the flank and keep Lee informed of the enemy's movements. Lee intended that Stuart should advance between the Army of Northern Virginia and the *Army of the Potomac* to provide cover for the entire army during the advance and to maintain contact with the enemy. Nevertheless, his orders did not specify the route which Stuart was to take. This permitted the latter to decide the matter for himself.

Stuart was rather sensitive about his reputation as a cavalryman and the unfortunate affair at Brandy Station seemed to cast a pall over that reputation. It—and the many skirmishes afterwards—seemed to demonstrate that the edge which his troopers had held over the Yankee cavalry was gone. He decided that something spectacular was needed to restore luster to his image. He had once ridden completely around the *Army of the Potomac*. Another such raid would serve both to repair his reputation and to provide useful information on the enemy's movements. Reasoning thusly, he decided to take three of his brigades and some horse artillery on a ride around the *Army of the Potomac*. He was confident that the move would be neither difficult nor dangerous. It would take perhaps four or five days to move the hundred or so miles through Virginia, across Maryland, and into Pennsylvania to reach Ewell in the Cumberland Valley. Moreover, Lee would still have available the services of three of Stuart's own brigades, those of Robertson, Jones, and Jenkins, plus Imboden's independent mounted brigade, which could be used for security and reconnaissance missions.

Stuart set out on 24 June, the very day when the lead elements of the *Army of the Potomac* began crossing the Potomac. He had with him the brigades of Wade Hampton, Fitzhugh Lee, and W.H.F. Lee, the latter under the command of Colonel John Chambliss, plus one battery of horse artillery. Immediately things began to go awry. Though a master at reconnaissance operating over familiar terrain, Stuart managed to chose the worst possible route around the *Army of the Potomac*. Failing to realize just how slowly the Yankees were moving, Stuart put his column on the very same roads which the *Army of the Potomac* was using. He became enmeshed in the Federal rear guards, fighting a series of small, though occasionally sharp, skirmishes. By 28 June he was finally clear of the Union rear guards. But at a time when he ought to have been well into central Pennsylvania, he had advanced no more than 35 miles and was actually only at Rockville, barely a dozen miles from Washington. There, while one brigade began tearing up railroad track, another overran a heavily ladened Federal supply train, taking 400 prisoners, and 125 wagons. Stuart paroled the prisoners the next day, but decided to take the wagons along rather than burn them. This further slowed his march. Nevertheless, by great efforts, he managed to advance 45 miles in the next two days, skirmishing with elements of the *1st Delaware Cavalry*, part of the Union *VIII Corps* assigned to guard railroads in Maryland, on 29 July. Late on the morning of 30 June he entered Hanover, where unbeknownst to him Brigadier General Judson Kilpatrick's Union *3rd Cavalry Division* had just arrived.

Shortly before 1000, the *13th Virginia Cavalry*, Stuart's advanced guard, ran into the rear of Kilpa-

trick's division. The surprised Union rear guards broke and fled. Chambliss' troopers drove right into the town, overrunning some Union ambulances, but Kilpatrick's *1st Brigade*, under the youthful Brigadier General Elon J. Farnsworth, rallied, and drove the Rebels off. Encumbered by the wagon train and with Fitzhugh Lee's brigade and Hampton's both at some remove from the main column, Stuart was unable to respond. He fell back, detouring around the town and advancing towards Carlisle, where he believed he would find Ewell, who was, however, by this time in the vicinity of Gettysburg.

Stuart reached Carlisle early on the morning of 1 July. There he found Union Brigadier General William F. Smith with a division of militia. He demanded that Smith surrender. When the latter refused, Stuart leisurely shelled his positions and put several buildings to the torch. A more serious engagement might have resulted, but during the action a messenger from Lee reached Stuart instructing him to return at once to the army, now concentrating in the vicinity of Gettysburg. Stuart immediately broke off the action and marched off. About noon the next day he brought his exhausted troopers back to Lee, who had sorely missed them. There was little to show for Stuart's week away from the army. To be sure, he had spread some disorder in the Federal rear and captured some useful booty, but his absence had hindered Lee's efforts to locate the enemy with the result that Lee was now engaged in a great battle under unfavorable circumstances. So seriously did Lee feel the loss of Stuart's services that the only occasion during the war on which he actually showed himself to be angry was on the morning of the third day of Gettysburg, when he was seen to raise his hand as if to strike Stuart. But all that, of course, was in the past. With Stuart present, Lee intended to make use of him. On 3 July, Stuart would strike at the Union rear from behind their right flank, even as Longstreet launched his great infantry attack against their center. Thus if the *Army of the Potomac* should break and be forced to retreat, Stuart would be ideally placed to exploit the victory.

By noon on 3 July Stuart was in position with four brigades several miles east of the Confederate left flank. There on the Rummel farm, his division went into action at about 1500, around the time Pickett's Charge was beginning. Stuart clashed with Union Brigadier General David Gregg's *2nd Cavalry Division*, reinforced to three brigades by the addition of Brigadier General George A. Custer's brigade of Michiganders from Kilpatrick's division. Gregg and his brigadiers were fully as tough and aggressive as were Stuart and his subordinates. A very confused engagement resulted. It began with an exchange of cannon fire and some skirmishing between dismounted troopers. In this the Union troopers seem to have gotten the better of their Rebel opponents, if only because many of them carried the Spencer repeating rifle, but the difference was slight. A more serious mounted action followed. Several charges and countercharges followed, in one of which Custer is reported to have cried, "Come on, you Wolverines!" as he gleefully plunged once more into the fray. There was much hand-to-hand fighting and at one time Gregg appears to have actually surrounded most of Stuart's men, only to have them cut their way out again. Tactically, it ended in a draw and each side claimed to have forced the other to give up the fight. However, the action should probably be labeled a Union victory in as much as Gregg frustrated Stuart's effort to disrupt the Federal rear. Losses were remarkably light, perhaps 250 for the Union and 200 for the Confederacy. The action had no effect whatsoever on the outcome of the great drama which unfolded simultaneously four miles to the southwest along Cemetery Ridge.

After Rummel Farm, Stuart and his cavalry returned to doing what they did best, serving as the eyes and shield of the Army of Northern Virginia as it retreated painfully back to its native soil. A good cavalryman, one of the greatest in history, Stuart had let his pride take him off on an unnecessary mission at a time when Lee's need for him had been greatest. In doing that, he had injured his reputation far more than Brandy Station had.

CHAPTER VII

Decision

3 July

*T*he sun rose strong and hot on Friday 3 July. Lee had spent much of the night planning an attack. Despite the poor results of the second day's fighting, the army was still determined and still full of fight. Unfortunately, rather than bring his principal subordinates together for a staff conference, so that they could express their views, Lee issued his instructions by courier. The plan was simple: he would simultaneously hit both ends of the Union line, collapse them, and force the *Army of the Potomac* to retreat. Ewell was ordered to reinforce his left under Johnson, with three brigades drawn from his right, under Rodes; he was then to attack the Union forces remaining on Culp's Hill early in the morning in an effort to exploit the gains Johnson had made after sunset the previous night. Meanwhile, Longstreet was to renew the attack against the Union left and center along Cemetery Ridge. And J.E.B. Stuart's cavalry, having finally turned up during the previous afternoon, was to undertake a diversion against the Union rear, threatening Meade's line of communications. Lee's plan was simple, but his instructions to his subordinates were neither clear nor forceful in detailing his desires. Longstreet, believing that the Cemetery Ridge line was much too strong for a frontal attack, began planning a flank march to the right in order to get around the Federal left beyond Big Round Top, which he believed to be but lightly held. At about dawn, soon

Major General George Edward Pickett who, despite graduating last in his class at West Point, managed to attain divisional command. His name would be forever linked to the disasterous charge against Cemetery Ridge on 3 July.

after Longstreet had issued orders designed to get his troops into position for such an attack, Lee rode up to his headquarters, curious as to why the attack had not yet begun. The resulting conversation between the two officers left Longstreet disappointed and depressed, and Lee stunned and frustrated. As it was too late to get an attack going in time to coincide with Ewell's, he had to improvise. Summoning up Longstreet and a few staff officers, he rode along Seminary Ridge, observing the Federal positions opposite. His eye fell upon a section of the Union front which looked particularly suited, a 650 yard wide stretch of almost bare ridge crowned by a prominent copse of trees, just at the north end of Cemetery Ridge before it rose to the higher ground of Cemetery Hill. There were about 1,300 yards between his skirmish line along the foot of Seminary Ridge and the Union front. Most of this was level, open fields crossed by the Emmitsburg Road and uncluttered save for a few farm buildings and fences. Lee explained to Longstreet, Hill, and the others present that he would smother the defenses with an artillery bombardment and then storm it with 15,000 men.

George E. Pickett

Born in Richmond, Pickett (1825-1875) graduated from West Point at the bottom of the class of 1846. An infantryman, he served in Mexico (one brevet) and on the frontier. While on garrison duty in Washington Territory in the period 1856-1861 he provoked a crisis with Great Britain when he occupied a disputed island in Juan de Fuca Strait. He resigned as a captain in late June of 1861 and was almost immediately commissioned a colonel in Confederate service. For several months he commanded the defenses along the lower Rappahannock and was named brigadier general in February of 1862. He led a brigade with notable elan during the Peninsula campaign until wounded at Gaines' Mill. He returned to command in time for the Antietam campaign, after which he was promoted to major general and given a division in Longstreet's First Corps. He led his division at Fredericksburg, in operations on the Virginia and Carolina coasts, and then during the Gettysburg campaign. On 3 July, three brigades from his division formed the right half of the Confederate assault against Cemetery Ridge which has ever after borne his name, Pickett's Charge.

After Gettysburg Pickett conducted operations against Federal enclaves on the Virginia and North Carolina coasts. Recalled to the Army of Northern Virginia during the campaign of 1864, he led his division in all the subsequent battles for Richmond, surrendering at Appomattox. After the war he declined an offer of command from the khedive of Egypt and of a U.S. marshalship from President U.S. Grant to enter the insurance business, in which he prospered. At the time of Gettysburg Pickett had already been married twice, once to an Indian princess who died in childbirth, and was engaged yet again to a teenaged girl who would survive him by 50 years. Pickett, who wore his hair in "long, perfumed ringlets that fell to his shoulders," was an able division commander but had the misfortune to have his name associated with a disastrous attack over which he had little control.

Longstreet vigorously expressed his opposition to the plan, but Lee would not be moved. As Henry Heth later put it, "The fact is, General Lee believed the Army of Northern Virginia, as it then existed, could accomplish anything." Longstreet was to attack as soon as practical with 11 brigades. Aside from Major General George E. Pickett's newly arrived division of Virginians of his own corps, most of the troops would come from Hill's Third Corps. These included all of Major

James J. Pettigrew

Pettigrew (1828-1863) was born into a planter family in North Carolina. An unusually bright young man, he graduated from the University of North Carolina at the age of 19 and was immediately appointed as an assistant professor at the Naval Observatory in Washington. He later spent several years studying law in both the United States and Europe, before setting up a practice in Charleston, South Carolina. Pettigrew was active in the state militia, and entered politics, being elected to the state legislature in 1856. In 1859 he served for a time as a volunteer officer with the French army in the Italian War, fighting at Solferino. Upon his return to Charleston he was named colonel of the 1st Regiment of Rifles, a Charleston militia outfit. During the secession winter, Pettigrew's regiment occupied several of the harbor forts and later took part in the bombardment of Fort Sumter. He then enlisted in the Hampton Legion, was shortly promoted colonel of the 12th North Carolina and was named brigadier general early in 1862. He commanded a brigade in the opening phases of the Peninsula campaign, during which he was seriously wounded and captured. Upon his exchange he commanded the defenses of Petersburg and later on the North Carolina coast.

During the Gettysburg campaign Pettigrew led a brigade in Heth's Division and assumed command when the latter was wounded. His division formed the left half of the column which attacked Cemetery Ridge on 3 July. Although wounded in the hand, Pettigrew remained on the field, directing the retreat after the assault failed. With Heth's recovery, he returned to his brigade, which he commanded in the rear guard action at Falling Waters, 14 July, where he was mortally wounded.

General William D. Pender's relatively fresh division of Carolinians and Georgians, now under the command of Major General Isaac Trimble, an elderly but game officer who had been acting as Lee's chief of engineers. In addition, there was Major General Henry Heth's division temporarily under Brigadier General James J. Pettigrew, which had been unengaged since its battering on the first day, plus a pair of brigades from Major General Richard Anderson's division. Lee apparently also intended that the balance of Longstreet's Corps, the divisions of Hood, and McLaws, were to support the blow in some fashion, probably in the form of a secondary attack further to the right, simultaneous with the main assault

Meade's headquarters, the Leister house, a small two-room cottage just behind Cemetery Ridge, at which Meade held a staff conference at about 2100 hours on the night of 2 July.

or soon after it. However, he was again unclear in his instructions, and Longstreet appears to have believed these troops were to be held back until significant success had been gained in the center. Thus, as the troops began to deploy for the attack, their commanders were still not in complete agreement as to what was to take place. And they were certainly not sure of its wisdom, for Longstreet still expressed reservations about the attack. These concerns do not seem to have troubled Lee. Although aware that the Union position was a strong one, he had complete confidence in his troops, and no doubt that they could take Cemetery Ridge, thereby splitting the *Army of the Potomac* in two and winning the battle.

Unlike Lee, Meade had convened a meeting of his senior officers soon after the fighting ended on 2 July. They met at about 2100 hours at the Leister House, conveniently located just behind Cemetery Ridge in the *II Corps* sector, roughly in the center of the Federal line. Into the front room of the house crowded Meade, Slocum, and Hancock, his wing commanders, seven regular or acting corps commanders (Newton, *I*; Gibbon, *II*; Birney, *III*; Sykes, *V*; Sedgwick, *VI*; Howard, *XI*; and Williams, *XII*), chief engineer Warren, and Major General Daniel Butterfield, the army's chief of staff. Looking in through a window of the tiny room, Lieutenant Frank Haskell, aide to Brigadier General John Gibbon, observed that "...some sat, some kept walking or standing, two lounged upon the bed, some were constantly smoking cigars." As the wounded and exhausted Warren slept in a corner, the officers exchanged views on the day's action and on their prospects for the morrow. There was some discussion of the state of the army's supplies, for there was but one day's rations left at hand, but it was concluded that these could be eked out with local resources. Dispirited by the disastrous battering which his corps had received, Birney expressed doubts about the possibility of holding much longer. Newton, a respected engineer, had reservations about the security of the army's flanks. Some officers debated the possibility of a general withdrawal to the Pipe Creek position, about a day's march to the southeast. However, Hancock, Howard, and Slocum were all in favor of fighting it out. Meade, who was somewhat concerned with the irregular character of the Union front, suggested that should a withdrawal become necessary a tactical one of about 2,000 yards, to a line running just east of Rock Creek, would be best. Meanwhile, Butterfield calculated that the *Army of the Potomac* had perhaps 58,000 infantrymen available, which was probably a bit high. Then, with Meade's consent, he polled Hancock, Slocum and the corps commanders, asking each three questions. The first, whether the army should remain in position, was unanimously answered in the affirmative. On the second question, whether the army should stand on the defensive or attack, the response was equally unanimous for the defensive. Finally, on the matter of how long the army should remain in position, the majority of

John Newton

Born into an old and prominent Virginia family, Newton (1822-1895) graduated West Point second in the class of 1842, which included Abner Doubleday, George Sykes, John Pope, Richard H. Anderson, Lafayette McLaws, and James Longstreet, among others who would become prominent in the Civil War. Save for participating in the so-called "Mormon War" of 1858, Newton's entire career was spent working as an engineer on various military and civil projects. In September of 1861 he was appointed a brigadier general of volunteers and assigned to the defenses of Washington.

During the Peninsula and Antietam campaigns Newton commanded a brigade in the VI Corps and then a division at Fredericksburg. Newton was one of the officers who personally expressed to Lincoln their reservations about Major General Ambrose E. Burnside, which precipitated the latter's resignation. He was promoted major general of volunteers in March of 1863 and led his division with great distinction during the Chancellorsville campaign. At Gettysburg, Meade named him to command the I Corps in place of Doubleday, who had proved wanting. Newton led the corps until the following spring. Afterwards, his commission as a major general not having been confirmed, he went West, commanding a division of the IV Corps in the Atlanta campaign and later held an administrative command in Florida.

At the end of the war Newton was brevetted major general of both volunteers and Regulars. He continued in the Army, rising to brigadier general and chief of engineers in 1884, working on a variety of projects, the most notable of which was the blasting of the Hell Gate reefs in New York's East River. This task involved over 125 tons of explosives; he was ably assisted by Major General Mansfield Lovell, a former Confederate. Retiring from the Army in 1886, he served as a railroad president and as commissioner of public works in New York City.

the officers recommended that Meade consider alternative courses of action if the enemy delayed attacking by more than a day. Although he was not bound by this poll, Meade was pleased with it, for he had already decided to stay and fight it out, and had expressed as much in a message dispatched to Henry Halleck shortly before the meeting. It then being about midnight, the meeting broke up. As it did, Meade turned to Gibbon of II Corps, saying, "If Lee attacks tomorrow, it will be on your front." When asked why he thought so, Meade

Jennie Wade

Jennie Wade was a 20-year-old resident of Gettysburg engaged to be married to Corporal Johnston H. Skelly of the *87th Pennsylvania*. Her sister had given birth with great difficulty on 28 June and Jennie was caring for her at her home on Baltimore Street, less than 50 yards north of Cemetery Hill. On 1 July Jennie, her sister, and the infant found themselves in the midst of the battle of Gettysburg. Protected by the sturdy brick walls of the house, they lived for three days in the midst of the greatest battle ever seen in this hemisphere. On the morning of 3 July, while Jennie was making bread, a Confederate musket ball smashed through a door on the north side of the house, pierced another into the kitchen, and struck Jennie in the back, killing her instantly, the only civilian casualty of the battle. Nor was the tragedy complete, for unbeknownst to Jennie, her fiance Corporal Skelly had been wounded and taken prisoner at Winchester on 13 May. Transferred to Virginia, he died in a hospital on 12 July.

replied, "Because he has made attacks on both our flanks and failed, and if he concludes to try again it will be on our center." Then the officers went out of the house and rode off into the dark.

Most of the Union corps commanders had little to do after the staff conference. They toured their lines, spoke quietly with some of their officers and men, and then turned in for a few hours sleep. Not so Major General Alpheus Williams, the acting commander of *XII Corps*. He had much to do before he could rest. He was concerned about the Rebel brigade which had secured a lodgment at the southern end of Culp's Hill on the previous night. Strongly ensconced in trenches and earthworks, Brigadier General George H. Steuart's men posed a threat to Union tenure of the entire hill and potentially could unhinge the whole right. With permission from Slocum and Meade, Williams planned to eject the Rebels in the morning and spent three hours preparing his attack. The enemy position was strong, with a stone wall as well as the earthworks covering it in front; its left flank was secured by some 700 yards of swampy ground stretching southwestwards from the junction of Rock Creek and Spangler's Spring. But it was not impregnable, for Steuart's men did not hold the crest

A view of the terrain on Culp's Hill.

of Culp's Hill. And Brigadier General George S. Greene's
heroic brigade of Geary's division still held the northern end
of the hill over on Steuart's right flank. So Steuart's brigade
held a salient in the middle of *XII Corps* subject to fire from
three sides. Williams took advantage of this. Finding some
excellent sites for his artillery, he proceeded to position 26
guns at ranges varying between 600 and 800 yards from the
enemy, in such a way that they could pound every part of
Steuart's position. These would open the attack with a 15-
minute bombardment. Then Brigadier General John Geary's
2nd Division would go in against Steuart's right while Williams' own division, the *1st*, now under Brigadier General
Thomas Ruger, would pin the enemy's attention by applying
steady pressure against his left with one brigade held out as

a reserve. At about 0300 hours on 3 July Williams was satisfied with his arrangements. Rolling himself up in a blanket, he lay down on a broad, flat rock beneath an apple tree to snatch a half-hour's sleep. By that time, other men were already stirring in anticipation of the day's fighting to come.

Major General Edward Johnson, commanding the Confederate forces in the vicinity of Culp's Hill, was also planning an attack for that morning. He had been reinforced by four brigades, Brigadier General James A. Walker's famed Stonewall Brigade from his own division plus the brigades of Colonel E.A. O'Neal, and Brigadier General Junius Daniel from Rodes' Division over on the right of Ewell's Second Corps' right, and Brigadier General William Smith's brigade from Jubal Early's division, which held the corps' center. With three brigades already in the line, this gave him seven brigades for an attack designed to secure Culp's Hill. He already had Colonel J.M. Williams' Louisianans holding the ground beneath the northeastern side of the hill on his right, and he had shifted Brigadier General John M. Jones' Virginians to cover his left. In his center was, of course, Steuart's mixed brigade of Virginians, North Carolinians, and some secessionist Marylanders holding a portion of the hill itself. Johnson placed some of the reinforcements to cover his exposed left flank, but put most of them behind Steuart, intending that they attack directly up Culp's Hill. Johnson's dispositions were unsound. The terrain was highly unfavorable to an attack, being mostly wooded slopes with rocky outcroppings, cliffs, and depressions. Nor was there any place suitable for artillery to be deployed in any numbers. Moreover, neither Johnson nor Ewell held out any forces as a reserve available to exploit success should the Yankees be driven from the hill. It would almost seem as though the attack was being staged because an attack had to be staged. Johnson planned to hit the Union lines at about 0500 hours. Alpheus Williams hit first.

Williams' 26 guns opened up at about 0430 hours. For 15 minutes they subjected the Rebel positions to heavy fire.

Then, as scheduled, they ceased firing. Geary's division was supposed to attack at this point, but the course of events becomes unclear. Both Geary and Ewell reported that a furious attack was made and repulsed with heavy losses. Yet neither Brigadier General George Greene's brigade nor Brigadier General Thomas L. Kane's incurred any serious losses in the course of the day's fighting. What actually appears to have happened is what Williams himself reported: that the cannonade had barely ceased when Johnson unleashed his own assault. It was a confused, ill-coordinated attack uphill against superior forces. Walker's famed Stonewall Brigade appears to have carried it out unsupported save for a large volume of rifle fire delivered by the other brigades. Kane's Pennsylvanians and Greene's New Yorkers poured a deadly fire into the attacking troops. Walker's men came on repeatedly in a series of regimental attacks, renewing the assault each time with a fresh regiment. Walker was a seasoned veteran, and his men were superb soldiers, but the Union position was too strong and too stoutly defended. The attack faltered. As Walker's troops streamed back down the slopes, Johnson's other brigades pressed forward, so that the entire front along Culp's Hill was soon enveloped in a furious fire fight between the well-entrenched Yankees at its top and the Rebels at its foot, sheltering behind trees and boulders. The fire fight continued for some time. Then, at about 0800, Johnson ordered in O'Neal's Alabamians. The entire brigade attacked in line, easily storming up the lower slopes with such furor that it frightened Geary. But then heavy artillery and rifle fire pinned O'Neal down and his men sought such shelter as they could find. At about 0900 Johnson threw in Walker's men once more to the right of O'Neal's pinned troops. Once again attacking with great skill and elan, Walker's men were once again beaten off. At about this time Slocum interfered in William's conduct of the battle. Believing that the withdrawal of Walker's men presaged a general Confederate retirement, he ordered an attack across the boggy meadow just east of Spangler's Spring. The order bypassed Williams and went directly to Ruger. Ruger re-

The 29th Pennsylvania in action at Culp's Hill on the morning of 3 July.

ferred it back to Slocum for clarification, and the latter permitted skirmishers to first probe the enemy position. Unfortunately, Ruger's courier garbled the oral instructions which he carried to Colonel Silas Colgrove, who was commanding Ruger's *3rd Brigade*. Against his better judgment, and without thinking to refer the matter back to Ruger, Colgrove immediately committed two regiments. At about 1000, 650 men of the *27th Indiana* and *2nd Massachusetts* charged right into the front of Jones' Brigade, which had been shifted there during the morning. The Rebels laid on a very

Confederate Brigadier General George H. Steuart's mixed brigade of Virginians, North Carolinians, and secessionist Marylanders attempt to storm Culp's Hill on the morning of 3 July.

heavy fire. The Hoosiers advanced about 200 yards and then wavered, falling back with over 30 percent casualties. The *2nd Massachusetts* was even more determined. Advancing nearly 400 yards, it seized a portion of the Confederate lines, holding on for perhaps 5 minutes before retiring reluctantly under heavy pressure both to front and flank. As it fell back, it made at least one counterattack to beat off pursuing Rebel troops. By the time the regiment reached the safety of Union lines it had lost nearly half its men in an unnecessary maneuver which did nothing to bolster the defense. Meanwhile, at 1000

Johnson ordered Steuart and Daniel to storm Culp's Hill, despite the protests of both commanders that the attempt was clearly suicidal. Steuart formed two lines facing at right angles to his former lines with Daniel to his right. The troops attacked "in a most gallant manner." Nevertheless, as soon as they emerged from the sheltering ridge they were subjected to a deadly cross fire. Rifle balls and canister raked the lines in what one officer observed was "the most fearful fire I ever encountered." Daniel's brigade had some modest success, getting to within 50 paces of the Union lines before going to ground. But Steuart's brigade fared badly. Its exposed left was unable to stand the fire and broke, fleeing to the rear. The right pressed on briefly, wavered and then fell back in turn, forcing Daniel's men to pull back as well. Johnson's attack had failed.

Williams now went over to the attack himself, ordering a general advance to recover the positions lost on the previous night and to press the enemy back to Rock Creek. This was done skillfully and swiftly with units all along the line helping to drive back the exhausted Rebels. It was shortly after 1100. Though artillery fire continued another hour, and skirmishers traded shots even longer, the fight for Culp's Hill was over.

Williams' defense had been masterful. He had skillfully rotated regiments into and out of action, drawing upon his reserves as needed, thus bringing the brigades of Colonel Charles Candy and Brigadier General Henry Lockwood to the relief of Greene and Kane. He had even drawn upon Brigadier General Alexander Shaler's brigade of *VI Corps* for assistance after it had been thoughtfully sent by Meade at about 0800. Throughout the action the general had retained good control over his units and during the fight had conducted limited counterattacks to brush Rebel skirmishers back from his front. It was a difficult defense, but well conducted. Of course, it had been marred by tragedy. There had been the foolish, unnecessary, and wasteful, if spectacular, heroism of the *2nd Massachusetts* and the *27th Indiana*. And the very nature of the position had resulted in many casualties from "friendly fire," as Union artillerymen were forced to shoot over the heads of the infantry in several places. Despite the great courage of the Confederate troops the Union right remained securely anchored.

While their comrades to the left drove the enemy off Culp's Hill, the Union troops on Cemetery Ridge spent the morning of 3 July putting their positions into a state of readiness. The Union front remained much as it had been at the end of the fighting on the previous night. On the extreme right there was some cavalry and a battery posted on Wolf's Hill. About 500 yards west of these was *XII Corps*, still relatively uninjured, facing eastwards on a front of about 2,200 yards along Culp's Hill and on the ground south and east of it across Rock Creek. Supported by a fairly fresh brigade from *VI Corps*, the corps amounted to perhaps 8,500 infantrymen. Wadsworth's battered division of *I Corps*, perhaps 2,500 riflemen in all, held some 500 yards along the northern face of Culp's Hill. To Wadsworth's left, holding about 2,000 yards of front on three sides of Cemetery Hill as it jutted northwards like a great prow, was Howard's unfortunate *XI Corps*, which with 6,000 infantry had little fight left. About 300 yards to Howard's left was the main Union line along Cemetery Ridge, facing westwards and running roughly southwards. There were about 5,500 infantrymen from two relatively unscathed divisions of *II Corps*, occupying some 700 yards of front. Doubleday's battered division of *I Corps* held the 200 yards to their left with about 3,000 men. To their left was Caldwell's depleted *1st Division* of *II Corps*, holding about 900 yards of front with the assistance of two fresh brigades from *VI Corps*, for a total of a little more than 4,000 rifles. To Caldwell's rear was the battered *III Corps*, amounting to little more than a division in strength, perhaps 5,000 effective troops. The extreme left of the Union line was held by the *V Corps*, which had been heavily engaged on the previous day but could still put over 8,000 men into a line that stretched more than 2,000 yards all the way down to and around Big Round Top. Beyond *V Corps*, there were two brigades of cavalry on picket. In reserve were over 7,500 fresh riflemen from *VI Corps*, with one division covering the army's left flank from a position behind *V Corps* and the other two divisions behind *III Corps*; Robinson's battered division of *I Corps*, with about 2,500 men at the southern end of Cemetery Hill, were able to support either *XI Corps* or *II Corps*. The entire front was liberally seasoned with artillery. There were about 32 guns on Culp's Hill, 26 towards the east with *XII Corps* and 6 more facing northwards with Wadsworth. Howard's *XI Corps* had about 55 on Cemetery Hill, 20 of which were directed northwards and rest westwards. Facing westwards along Cemetery Ridge *II Corps* had 26 pieces of artillery, and there were 33 from the *Artillery Reserve* supporting Caldwell. *V Corps* had 12 guns in the line, 6 of which were on Little Round Top, and about 18 in reserve. Also in reserve were 48 guns with *VI Corps*, 20 with *III Corps*, and about 50 more from the *Artillery Reserve*. Altogether the *Army of the Potomac* had available perhaps 54,000 infantry and about 280 pieces of artillery—with some 6,500 artillerymen—to cover a front of over 9,000 yards. Thus, including reserves there were about 6 rifles per yard of front and about 1 gun for every 200 men, a comfortable average, although the line was rather thin in some areas, notably in Caldwell's sector. Most of the Union troops were under cover, even if only the shelter of a fence or a hastily dug trench, with their positions selected to give maximum fields of fire. Where the shape of the ridge permitted, there were two lines of infantry, one able to deliver fire over the heads of the other.

The Union position had some flaws, but it was fairly sound. By 1100 Meade's preparations were complete. He had toured the front, and been generally satisfied with the dispositions of the men and guns. In addition, he had issued orders to two important subordinates who were not present on the field. Major General William H. French commanded about 11,000 troops at Frederick and on Maryland Heights above Harper's Ferry, and Major General Darius Couch commanded the 32,000 militia and home guards along the Susquehanna. In the event of a victory, they were to bring pressure on the retreating enemy, French by striking against Lee's line of

John Gibbon

A native Philadelphian, Gibbon (1827-1896) was raised in North Carolina from boyhood. Graduating from West Point 20th in the class of 1847, along with Ambrose P. Hill and Ambrose E. Burnside, he served in Mexico and against the Seminoles, before putting in five years as an artillery instructor and quartermaster at West Point, finding time to write *The Artillerist's Manual*. His work was adopted as an official text by the War Department, and was used as a handbook by both sides during the war. Although his three brothers all entered Confederate service, Gibbon remained loyal to his oath at the outbreak of the Civil War. He served in various artillery posts for a time, but, since promotions in the artillery were slow, transferred to the infantry. Commissioned a brigadier general in the spring of 1862 he commanded the famed *Iron Brigade* in the *I Corps* during the Second Bull Run and Antietam campaigns. Later that year he was given a division, which he led at Fredericksburg, where he was severely wounded. In the spring of 1863 he returned to duty as a division commander in Major General Winfield Scott Hancock's *II Corps*,

which he led at Chancellorsville. During the Gettysburg campaign Gibbon led his division, and the corps when Hancock was assigned higher command, with great ability. It was against his front that the "High Tide" of the Confederacy washed.

Severely wounded at Gettysburg, Gibbon was placed on administrative duty until the following spring, when he resumed command of his old division. Gibbon fought in all the battles of the *Army of the Potomac* from the Wilderness to Petersburg, earning a major generalcy in the process. In early 1865 he was given the *XXIV Corps* in the *Army of the James*, which he led until Appomattox, when he was named one of the commissioners to receive the surrender of the Army of Northern Virginia. After the war he became a colonel in the Regular Army. Commanding the *7th Infantry*, he saw service on the frontier for the next 30 years, by which time he was a brigadier general in the Regular Army. Gibbon's memoirs, written in 1885, were not published until 1928. He was one of the ablest division commanders of the war.

retreat in the Shenandoah Valley and Couch by advancing southwards and coordinating his movements with those of the *Army of the Potomac*. Should the battle turn out badly, both officers were to fall back, French through Frederick and thence on to Washington and Couch back across the Susquehanna to cover central Pennsylvania. Meade broke for lunch, accepting the invitation of John Gibbon of *II Corps* to share some stewed chicken with Hancock, Newton, and Pleasonton.

Not far from where Meade and Gibbon lunched, there was a large stand of timber known as Zeigler's Grove on the right flank of the *II Corps'* front. About 50 yards to the southwest of this was the start of a low stone fence in rather poor repair with a rail fence built over it in places. This ran roughly southwards for about 200 yards before turning 90 degrees to the right to run west for nearly 100 yards more; it then made yet another 90 degree turn to the left to run southwards once more and end about 125 yards further on. Sheltering right behind the angle formed by this wall, directly in the middle of the corps' line, was 28-year-old Brigadier General Alexander Webb and his brigade of Philadelphians from Brigadier General John Gibbon's *1st Division*. They were supported by 22-year-old Lieutenant Alonzo H. Cushing's *A Battery, 4th Artillery*, a Regular Army outfit with six 3-inch rifled cannon. Crowning the almost bare ridge at this point was a very prominent copse of trees. It was these which had attracted Lee's eye while on reconnaissance that morning, and these towards which he intended to direct his attack.

Lee's army spent the morning preparing for the attack. Colonel Edward P. 'Alexander, ablest of the artillerymen in First Corps, directed the emplacement of the guns which would fire the preliminary bombardment, assisted by Colonel R. Lindsey Walker and Colonel J. Thompson Brown, who commanded the artillery in Third Corps and Second Corps. The guns, numbering around 150, were deployed on a 7,500-yard front stretching from Oak Hill—where two very long range Whitworth cannon were located—to the Peach Orchard. Most of them were placed in three grand batteries. About half, 75, were deployed on a 1,300-yard front stretching from the Peach Orchard roughly northwards along Seminary Ridge, and another 35 were similarly posted on an 800-yard front about 300 yards further north along the ridge; yet another 26 were about 300 yards further along the ridge and the balance were scattered at odd sites all along the front.

How Many Guns Bombarded Cemetery Ridge?

The number of Confederate cannon taking part in the bombardment of Cemetery Ridge on 3 July is given variously as anything from 138 to 179 in accounts which are all more or less equally reliable. The figure used here is "about 150," which is basically that of Lee's chief of artillery, Brigadier General William N. Pendleton, who said "nearly 150." This can be reconciled rather readily with the highest number given, 179, cited by Colonel Edward P. Alexander, who was responsible not merely for the guns which were to conduct the bombardment, but also for about two dozen short range howitzers which were to be held in reserve and used to accompany the infantry attack. Merely deducting these from the figure of 179 brings it down into the "about 150" range. Neither is it any more difficult to reconcile this figure with the lowest estimate, 138, which is clearly too low. At least 142 guns can be definitely demonstrated to have fired during the bombardment, which is certainly within the limits of Pendleton's "nearly 150." As result of these calculations we are left with between 142 and 155 guns. Since it is impossible to clarify the matter further, the figure "about 150" appears reasonable. On the other side of the battlefield were apparently 106 guns deployed along Cemetery Ridge. However, only about 85 actually engaged in counter-battery fire during the bombardment, and most of these for short periods only, as Union chief of artillery Brigadier General Henry Hunt preferred to save ammunition for the expected infantry attack. There was room along the ridge for many more guns, but 136 or so were being held in reserve for use once Confederate intentions became clearer.

Although the number of pieces was impressive, in fact many of the guns were poorly positioned, being much too far from the main objective for their fire really to be effective. Nevertheless, Brigadier General William N. Pendleton, Lee's chief of artillery and an ordained Episcopal minister, approved all of Alexander's dispositions. About two dozen howitzers were held in reserve to be moved forward when the preliminary bombardment was over, lending close artillery support to the attacking infantry. Much to everyone's surprise, the Union artillery made no effort to disrupt the preparations, which were readily visible from Cemetery Ridge, for Brigadier General Henry Hunt had decided to conserve ammunition for

Confederate artillerymen at rest, from an engraving made from a wartime sketch.

use in the forthcoming attack. As the guns moved into position, the infantry also began to get ready.

The attacking infantrymen deployed on a front of about 1,350 yards behind the western side of heavily forested Seminary Ridge. Their lines were drawn somewhat obliquely to the Union front, roughly northeast-southwest rather than north-south. When they advanced, the troops would have to correct this. Major General George E. Pickett's fresh division of Virginians was on the right, deployed in two waves on a front of about 825 yards, each wave consisting of two ranks with a thin line of file closers behind. On the right were 1,600 men of Brigadier General James L. Kemper's brigade, resting on the Peach Orchard. To its left was Brigadier General Richard B. Garnett with 1,500 more men. About 100 yards behind one of these brigades—it is unclear which—was Pickett's second wave, the large brigade of Brigadier General Lewis A. Armistead, over 2,000 men on a front of about 500 yards.

About 325 yards to Armistead's left was Heth's Division under Pettigrew, which had been badly battered on the first day but had been unengaged since. Pettigrew's men were also deployed in two waves, but rather differently. He had four brigades in line on a front of about 700 yards. Each brigade had half of each regiment deployed in the first wave and the other half in the second wave about 100 yards to the rear. On the right were the Tennesseans and Alabamians of Brigadier General James J. Archer's brigade, now under Colonel B.D. Fry, perhaps 500 men. Then came the North Carolinians of Pettigrew's own brigade, some 1,200 men under Colonel J.K. Marshall, with Brigadier General Joseph R. Davis' 1,200 Mississippians and North Carolinians to their left, and the 500 Virginians of Colonel J.M. Brockenbrough's brigade, under Colonel Robert M. Mayo. About 100 yards behind Pettigrew's troops was a third wave, consisting of two small brigades of North Carolinians from Major General William D. Pender's division under Major General Isaac Trimble, with Brigadier General Alfred Moore Scales' 1,200 men on the right and Brigadier General James H. Lane's 1,700 on the left along a front of about 550 yards. In addition, there were two brigades from Major General Richard Anderson's division deployed on a 400-yard front some 250 yards to the front and right of Pickett's Division. These troops, Brigadier General Cadmus A. Wilcox's 1,200 Alabamians on the right and Colonel David Lang's 700 Floridians, were to cover the exposed right flank of the attacking brigades, thus permitting them to move forward. Quite surprisingly, no such arrangements were made to protect the equally exposed left flank of the attacking brigades. In addition to the nine brigades actually making the attack plus the two covering the right

Meteorology

The Gettysburg campaign took place during the last weeks of spring and the first weeks of summer, traditionally the ideal campaigning season with generally fine weather and long days.

Weather

For the early part of the campaign the weather was generally seasonable. It was rather hot until 25 June, when there was a cooling trend. For the critical days of the campaign there is fairly detailed information as a result of the efforts of the Rev. Dr. M. Jacobs of Pennsylvania—Gettysburg—College and the more casual observations of Lieutenant Frank Haskell.

With the rains on 4 July the temperature fell sharply. Rains on the morning of 5 July helped cover Lee's retreat from the field. The next day was clear, but the rains resumed on the 7th, and grew quite heavy on the 8th, by which time Lee had reached the security of the Antietam area.

Note that during the battle the temperature was at its highest at 1400 on 3 July during the Confederate bombardment preceding Pickett's Charge.

Date	Day	Temperature (F)			Precipitation
		0700	1400	2100	
25 June	Thu	59	51	63	Intermittent rains
26 June	Fri	60	63	62	from 0800 25 June
27 June	Sat	61	63	67	to 0700 27 June.
28 June	Sun	63	67	68	
29 June	Mon	66	72	69	
30 June	Tue	68	79	71	
1 July	Wed	72	76	74	Very cloudy.
2 July	Thu	74	81	76	Foggy dawn; humid; PM, "Mizzling rain."
3 July	Fri	73	87	76	Very humid; thunderstorms nearby.
4 July	Sat	69	72	70	Very humid; intermittent showers, 0600-1200; heavier rains, 1300-1400; very heavy rains from 2100 on.

Sunrise/Sunset

Local sunrise and sunset were the principal determinants of when fighting could take place, since engagements at night were risky at best. Gettysburg lies at 39° 50' North and 77° 16' West, which yields the times for sunrise and sunset as listed below.

Under moonless conditions it is theoretically possible to engage in military operations from about the beginning of morning nautical twilight. This occurs some 30 minutes earlier than the beginning of civil twilight, and is essentially when visibility is about 400 yards under ideal conditions. Operations can continue until the end of evening nautical twilight, about 30 minutes after the end of civil twilight. In practice, of course, conditions are rarely sufficiently ideal for operations.

	30 June	1 July	2 July	3 July	4 July
Civil Twilight Begins	0403	0403	0404	0404	0405
Sunrise	0435	0435	0436	0436	0437
Sunset	1933	1933	1933	1932	1932
Civil Twilight Ends	2006	2006	2006	2005	2005

The Moon

By coincidence, the Gettysburg campaign lasted almost exactly two lunar months. The Army of Northern Virginia began to march north in early June by the light of the full moon. By the time Ewell's Second Corps reached Winchester there was virtually no moon, which may have helped his initial night attack on the place. As the Confederate army began to spread out and plunder southeastern Pennsylvania the moon began to wax, reaching full just before the armies clashed at Gettysburg. The cloudy conditions and fog on the night of 1-2 July largely obscured the moon, but it appears to have been fairly bright on the night of 2-3 July with sufficient light to cause Ewell concern about continuing the advance of Johnson's Division on Culp's Hill; there was certainly enough moonlight to permit Alpheus Williams to site the batteries for his planned morning attack against Johnson's positions. There was little moonlight during the retreat.

Phases of the Moon:	June	July
Full Moon	1	
Last Quarter	7	7
New Moon	15	15
First Quarter	23	23
Full Moon	30	31

Confederate infantrymen awaiting the end of the bombardment of Cemetery Ridge and orders to launch the great assault on the Army of the Potomac.

flank, three other brigades of Anderson's Division, Brigadier General Ambrose Wright's Georgians, Brigadier General William Mahone's Viginians, and Brigadier General Carnot Posey's Mississippians, were available and were told to hold themselves ready to support the attack on orders. In the attacking brigades, each front line regiment had about 10 percent or more of its men out as skirmishers. These had filtered into the area between the two armies to get within rifle shot of the Union lines. During the morning they had helped pull down fences and other obstacles which might hamper the orderly advance of the attacking brigades. Now they lay quietly under cover, awaiting the attack. When it came, they would open a harassing fire on Cemetery Ridge and then advance in two lines, the first about 100 yards in front of the main line and the other 100 yards in front of that. Just before the attack struck the Union lines they would be reabsorbed into their brigades. Preparations for the attack appear to have been largely completed by about 1100. The troops rested as best they could as the temperature rose into the high 80s. Some men napped, others had lunch. Wilcox, who was commanding one of the flank guard brigades, invited Garnett, who was to lead one of the attacking brigades, to lunch with one of Pickett's staff officers. It was a simple meal of tepid mutton eked out with some very hard and very cold well water liberally doused with some very good whiskey liberated from a local farm house.

All through that hot morning there had been firing along the Cemetery Ridge position. Some of it was just casual shooting by bored soldiers or sniping by skirmishers posted in advance of the front by both sides. But there were also some artillery rounds fired, mostly for ranging purposes, and there had been one or two small artillery exchanges. At one point the *1st Company, Massachusetts Sharpshooters* went into this no-man's-land to clear away Rebel snipers. At about 1000 a considerable skirmish occurred at Bliss' barn, some 400 yards from the Union lines, when elements of the *12th New Jersey* ejected a number of Confederate skirmishers who were sniping at Cemetery Ridge. The Jerseyites no sooner withdrew, however, than the snipers returned. The *14th Connecticut* went in, retook the place, and burned it to the ground to the accompaniment of an exchange of artillery. By about 1100 the skirmishing had come to an end, and only distant sounds of firing around Culp's Hill disturbed the air. That ended at about noon. An odd silence descended over the battlefield. On both sides the troops were remarkably confident. Their officers largely shared this sentiment, save for Longstreet himself who grew nervous. He remained unhappy about the

A Matter of Time

Time presents a difficult problem when dealing with any pre-20th century event. Timepieces were not reliable, nor did standard time exist before 1884. As a result, all time was local solar time. In the case of the Gettysburg campaign the times given for particular events by various participants frequently differ by as much as 30 minutes. A good case in point is the Confederate bombardment of Cemetery Ridge on 3 July. Colonel J.B. Walton, chief of artillery of the Confederate Second Corps, recorded receiving Longstreet's order to fire at 1330. Since he immediately ordered the signal shots to be fired, by his reckoning the bombardment could have begun no later than 1335. On the other hand, Colonel Edward P. Alexander, who was actually in charge of the bombardment, reported fire commenced at almost exactly 1300. And in Gettysburg, the Reverend Dr. M. Jacobs, professor of mathematics and chemistry at Pennsylvania College, an inveterate note taker, recorded that it began at 1307 hours. Similar discrepancies exist in observations on the duration of the bombardment and the time of its end. Professor Jacobs, who recorded that the bombardment ended just before 1500, is probably to be most relied upon, at least for the bombardment, since he was after all a local resident. For other events the matter is less easily settled. While most firsthand reports agree that the attack on Cemetery Ridge on 3 July required about 20 minutes, there is little agreement on anything else. All times cited in this account have been based as much as possible on a consensus of available reports, modified by knowledge of the military practice of the period, and the inherent time required for various activities to occur. Thus, all times given are essentially approximations.

---

entire enterprise and at one point even attempted to delegate to Alexander the authority to call off the attack, writing, "If the artillery does not have the effect to drive off the enemy or greatly demoralize him so as to make our effort pretty certain, I would prefer that you should not advise General Pickett to make the charge." A remarkable order for a corps commander to give to a colonel. Alexander penned a brief note pointing out that if there remained any doubts to its wisdom, the attack should be called off before the bombardment began, for it would consume most of the remaining artillery ammunition. This seems to have brought Longstreet back to his senses, for his next message more reasonably instructed Alexander to notify Pickett when he thought the artillery had done its work, to which Alexander replied: "When our Arty. fire is at its best I will advise Gen. Pickett to advance." Soon after, reluctantly satisfied that all was in readiness, Longstreet passed a note to Colonel J.B. Walton, his chief of artillery, "Let the batteries open." Walton passed the order to fire the agreed upon two signal rounds. At 1300 hours Captain M.B. Miller's 3rd Company of Louisiana's Washington Artillery Battalion fired a single shot from a 12-pounder Napoleon. A second gun misfired, but a third fired within seconds, giving the requisite second shot. Almost instantly a wave of flame and smoke leaped from the nearby batteries and spread rapidly along the line northwards; some 150 guns had begun the most intensive bombardment ever seen in the Americas.

As shells and solid shot rained down, the Union front became alive with activity. Infantrymen sought cover, as artillerymen stood to their pieces. Henry Hunt, chief of artillery on Little Round Top, tarried briefly to observe the scene, later recalling that it was "indescribably grand," and then rode off to see about bringing up more guns and more ammunition. Lingering over the remnants of lunch, Gibbon leaped up, grabbing his sword, and shouting for his orderly to bring his horse. As the man led the mount up, he was struck in the breast by a shell fragment and killed instantly. The horse fled. Gibbon headed for the front on foot. The imperturbable Hancock continued dictating an order concerning fresh beef for *II Corps*, the command of which he had just resumed from Gibbon. When he finished, he walked up to where Gibbon stood to engage the latter, a gunner himself, in a professional discussion about the effectiveness and purpose of the bombardment. Gibbon was of the opinion that it presaged a retreat—Hancock felt certain it was preparatory to an attack. Meanwhile, the rain of shot and shell continued. Chaplain Alanson A. Haines of the *15th New Jersey*, a *VI Corps* outfit posted in reserve, later wrote, "...a terrific rain of hundreds of tons of iron missiles were hurled through the air. The forests crashed and the rocks were rent under the terrible hail...the smoke was impenetrable, and rolled over the scene

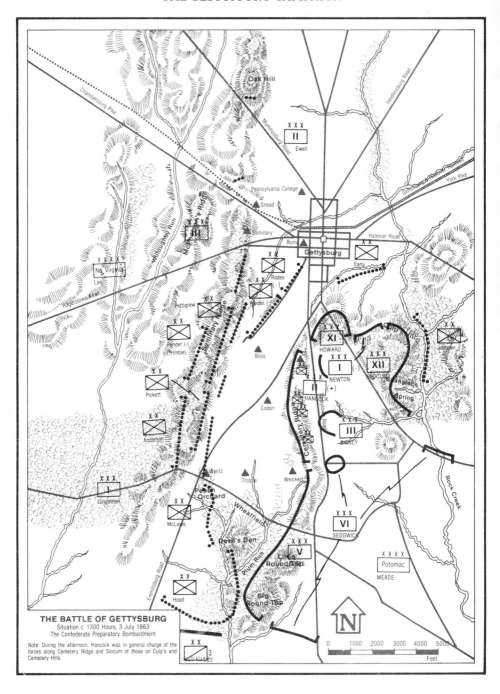

THE BATTLE OF GETTYSBURG
Situation c. 1300 Hours, 3 July 1863
The Confederate Preparatory Bombardment

Note: During the afternoon, Hancock was in general charge of the forces along Cemetery Ridge and Slocum of those on Culp's and Cemetery Hills.

Winfield Scott Hancock

A native Pennsylvanian, Hancock (1824-1886) graduated from West Point 18th in the class of 1844. He served on the frontier, in Mexico (one brevet), against the Seminoles, in Kansas, and in the "Mormon War." The outbreak of the Civil War found him post quartermaster at Los Angeles, where he presided at a famous dinner for several officers who had resigned to "go South." He was appointed a brigadier general of volunteers in September of 1861 and led a brigade in *IV Corps* through the Peninsula campaign. At Antietam he commanded a division of *II Corps*, and, after promotion to major general, at Fredericksburg, and Chancellorsville, where he particularly distinguished himself in covering the retreat of the *Army of the Potomac*. Given command of the corps in mid-May of 1863, he led it into Gettysburg. There Meade, a close personal friend, assigned him to direct the battle on the afternoon of 1 July until the latter could arrive. Hancock is largely responsible for preventing the defeat which the Army of Northern Virginia had inflicted upon *I Corps* and *XI Corps* earlier that day from turning into a disaster. On 2 July he commanded much of the left wing, endeavoring to salvage something from the disas-trous situation which developed upon the collapse of *III Corps*. It was on 3 July, however, that he proved particularly capable, commanding most of the forces on Cemetery Ridge, and by a brilliant display of courage and leadership helped break the Confederate grand assault that afternoon, during which he was wounded.

After Gettysburg, Hancock resumed command of his corps, leading it in most of the engagements of the campaign of 1864 until November, when his Gettysburg wound reopened. No other Federal officer commanded a corps for a longer period during the war. He thereafter directed various administrative commands until the end of the war. After the war Hancock, a Regular major general, served in a variety of posts. In 1880 he ran for president as a Democrat, and was beaten by fellow-veteran James A. Garfield by less than 10,000 popular votes though the Electoral vote was 214 to 155. He returned to active service and died in command of the Department of the East, at Governor's Island, New York. One of the finest soldiers of the war, Hancock had an imposing presence, a blistering vocabulary, and a talent for inspiring men under fire.

---

of action, concealing all...." Most Union batteries withheld their fire as ordered, though a few in *XI Corps* and those atop Little Round Top were permitted to open a selective reply,

Commander of the II Corps, Major General Winfield Scott Hancock. Hancock was one of the best Union corps commanders who proved himself in many of the greatest battles of the war. At Gettysburg, where he played a major role throughout the battle, he suffered a wound while leading the defense of Cemetery Ridge on 3 July.

attempting to find targets through the smoke which shrouded the enemy's guns.

The barrage enveloped the entire Union front from Cemetery Hill down to the end of Cemetery Ridge. Men and horses were killed and wounded. Guns were disabled. Ammunition wagons were struck, disappearing in huge explosions, which elicited great cheers from the Rebel ranks. But after the opening rounds Confederate fire tended to be a little high so that many shots missed the ridge entirely and fell behind it. Several landed among the wagons of the *Artillery Reserve*, forcing the commander Brigadier General Robert O. Tyler to pull it back about a half mile. Other overages disrupted troop concentrations and supply wagons, and panicked rear-area personnel and sutlers. As Meade's headquarters were behind Cemetery Ridge at the Leister house, it took a number of rounds. Butterfield, the chief of staff, was wounded, other officers had close calls, and 16 of the staff's horses were killed. Meade stood fast through most of the bombardment, prefer-

Alexander S. Webb

Born into a distinguished New York City family—his father was the noted diplomat and newspaperman James Watson Webb—Webb (1835-1911) was privately educated and graduated from West Point 13th in the class of 1855. An artilleryman, his career encompassed duty against the Seminoles, on the frontier, in garrison, and as a mathematics instructor at West Point. At the start of the war he was in garrison at Fort Pickens, near Pensacola in Florida. As a captain, and later major, he served as an aide to the chief of artillery at Bull Run and in all the campaigns of the *Army of the Potomac* from the Peninsula to Antietam, during which he became chief of staff of the *V Corps*. He then commanded a training camp in Washington, before returning to the *V Corps* as assistant inspector general during the Chancellorsville campaign. Promoted to brigadier general of volunteers, he was given the *2nd Brigade* of Gibbon's *2nd Division* in Hancock's *II Corps*, which he led with considerable panache during the Gettysburg campaign, holding the Angle atop Cemetery Ridge on 3 July against which the "High Tide of the Confederacy" broke. He was wounded and years later he was awarded a Medal of Honor.

Webb commanded the division during the recuperation of Gibbon in the autumn of 1863 and winter of 1864, when he returned to his old brigade. He led it in the Wilderness and Spotsylvania, where he was so severely wounded as to be unfit for duty until January of 1865, when he became chief of staff to Meade. He ended the war as a brigadier general of volunteers, brevet brigadier general of Regulars, and brevet major general of both Regulars and volunteers. In 1866 Webb was made colonel of the *44th Infantry*, and then taught at West Point once more. In 1869 he was named president of the City College of New York. He requested and received an honorable discharge, and served C.C.N.Y. for 33 years.

---

ring to be at the post of danger to changing his headquarters and thereby possibly disrupting communications. However, he later wandered over to Powell's Hill for a while to confer with Slocum, thereby missing most of the subsequent events. Dangerous as it was behind Cemetery Ridge, it was hotter still on the front itself. And nowhere was it hotter than on the lines of *II Corps*, where the fire of over 100 guns was concentrated. Cushing's battery alone lost half its guns, three ammunition chests—an eighth of its supply—and several men with them. The other batteries and the infantry along the

Brigadier General Alexander S. Webb. His brigade of four regiments occupied the Angle during Pickett's Charge.

stone wall suffered as well. At about 1315 Hancock ordered the batteries to reply as much to bolster morale as to inflict damage upon the enemy. Five batteries opened up with 25 guns, much to Hunt's annoyance, but with salutary effect on the troops. The Rebel gunners slowed their rate of fire, coordinating it in battery volleys for better accuracy and to maintain a relatively high sustained fire. The Union batteries on Cemetery Hill deliberately fired intermittently, convincing the enemy that many guns had been put out of action. The infantrymen took the pounding well, clinging to the trembling earth in the broiling sun and yelling at each other over the thunderous roar, as their officers walked back and forth along the lines, shouting encouragement. Casualties among them were relatively light, due partially to the rudimentary shelter behind which most of them lay and to the fact that many Rebel shells were duds, but mostly to the fact that overages tended to increase as time passed, a phenomenon of the necessity of relaying the guns after each shot. Indeed, the number of shells falling behind the lines was so great as to discourage the fainthearted from fleeing. It seemed safer at the front than behind it.

Some units suffered greatly. The 25 guns in front of *II Corps* seem to have been reduced to no more than half that number by the shelling, and several had to be pulled back. Two regiments, the *108th* and *126th New York*, posted behind Lieutenant George A. Woodruff's *I Battery, 1st Artillery*, a Regular Army outfit with six 12-pounder Napoleons, discovered that Zeigler's Grove offered little protection. Shells bursting among the trees sent wood splinters in all directions, killing or wounding many. A brigade of *III Corps* had the misfortune to be the target of a number of explosive shells and suffered severe losses. Psychologically, of course, the question of whether the bombardment was or was not accurate was unimportant. The troops were being pounded, probably in preparation for a grand assault. No man knew when a shell might seek him out, and when the shelling did stop it would only mean that they would soon be locked in a titanic death struggle with the enemy's infantry.

The Rebel infantry was being pounded as well, albeit more lightly, by such Union guns as were permitted to return fire. Though most of the troops were under cover, casualties were taken in some cases in substantial numbers. The most exposed brigade, Brigadier General James J. Kemper's, suffered a loss of about 15 percent and Pickett's Division took perhaps 300 casualties overall. For the Confederate troops the shelling was doubly terrible, for it presaged the pounding they knew they would be subjected to as they advanced across the more than 1,300 yards of open ground which separated them from their objective. But the men had confidence in their officers and almost literally worshiped Lee. They stood their ground and steeled themselves for the attack, much as their foes opposite did in anticipation of it. To inspire and calm them, their commanders passed among them, ignoring shot and shell.

On Cemetery Ridge senior officers also deliberately, often recklessly, exposed themselves in order to steady the men.

Brigadier General Alexander Hays, commanding the right hand division of *II Corps*, ran exuberantly from place to place behind the lines, shouting out boyish slogans, laughing, and cheering for the troops. Gibbon spent part of the bombardment sitting quietly with his aide, Lieutenant Frank Haskell, just behind the crest of Cemetery Ridge, until the two of them went for a stroll along the lines, pausing now and then to chat with the infantrymen crouching behind whatever shelter they could find. Then they descended down the west face of the slope to the shade of some elms to sit with a magnificent view of the enemy batteries blazing away over on Seminary Ridge. Brigadier General William Harrow, commanding Gibbon's left hand brigade, walked quietly up and down the lines with his arms folded, coolly ignoring the shower of fire. Brigadier General Alexander Webb, a New York aristocrat commanding a bunch of Irish roughs from Philadelphia, stood quietly in an exposed position, leaning on his sword, calmly puffing a cigar. Doubleday sat on the ridge eating lunch; when a shell threw some gravel on his sandwich he cracked, "That sandwich will need no salt!" Howard sat down with his staff on the forward slope of Cemetery Hill beneath one of his batteries to watch the spectacle. And Hancock appeared everywhere, neatly uniformed and riding a magnificent black horse. At about 1400 he trotted down to McGilvery's silent batteries on a rise just south of Cemetery Ridge, and ordered them into action. McGilvery reminded him of Hunt's orders that he withhold fire, but obeyed. Over 80 guns were now replying in some fashion to the Confederate barrage. Hunt, who was also all over the field that afternoon, was incensed. He had been busy arranging for the movement of additional batteries and ammunition, and wanted no interference in his preparations. Furthermore, the artilleryman wished to conserve ammunition for the expected infantry assault. Having consulted with several other officers, including *XI Corps* chief of artillery, Major Thomas W. Osborn, he had concluded such counter-battery fire as the Union guns were already delivering had been relatively ineffective. He rode over to find Meade. Meanwhile, Gouverneur Warren had been back on

*A view of the ground over which Pickett's Charge was conducted,
showing not only its gently rising nature, but also the stone fences
which helped define the Union line.*

Little Round Top carefully observing the fall of shot on the
enemy's lines. He too concluded that the counter-battery fire
was ineffective and informed Meade. Meade decided it was
better to conserve ammunition. Although Hunt failed to find
the general commanding, he soon received orders to cease
firing and issued the necessary instructions. As the batteries
were widely dispersed, they received their orders at different
times. Thus Union artillery fire did not cease abruptly, as if by
design, but rather died down gradually, as if the gunners
were running out of ammunition. By 1445 the Federal guns
were silent.

In the Confederate lines, Colonel Alexander had been
carefully observing the effects of the bombardment. He
believed that the nearly two hours of shelling had done their
work. He had noted the ragged way the Union counter-fire
had slackened and then ceased and he had also noted that on
the *II Corps* front some batteries seemed to have been pulled
out of the line. He concluded that Yankees were running out
of ammunition and had stopped firing to conserve what they
had for the expected infantry attack. As his own batteries

were themselves getting low on ammunition, he decided that the time was ripe. He hastily penned a short note to Pickett, beginning, "For God's sake come quick," and followed it with two verbal messages. Then, as the Confederate guns fell silent, he waited impatiently. Alexander's written message found Pickett in conversation with Longstreet. Upon reading it, he passed it to the latter. Longstreet sat silently, message in hand. Pickett, who had twice asked Alexander if the time was ripe, spoke up, "General, shall I advance?" Unwilling to speak, Longstreet bowed his head. Pickett saluted, saying "I shall lead my division forward, sir," then passed Longstreet a note for his fiancee and rode off to his troops. Longstreet rode over to Alexander's post. There he discovered that the reserve guns were nowhere to be found—Pendleton had moved them to the rear and neglected to inform Alexander. This meant that the attacking infantry would have to be accompanied by batteries which had participated in the bombardment, batteries for which there was little ammunition and which had tired gunners. He suggested halting the advance until the guns could be resupplied. Alexander informed him that it would take at least an hour to do so. Though stunned and now very dubious about the wisdom of the undertaking, Longstreet had no choice—the attack had to go forward immediately.

The ranks of the attacking brigades stirred. Orders were issued. The troops prepared for battle each after his own fashion. In some units they joined their officers in prayer. In others they listened, attentively or not, as their commanders exhorted them with patriotic slogans or heartened them with folksy comments. A couple of generals conferred together briefly. Some officers assigned reliable men to keep an eye on those less so. It required but a few minutes to get the men ready, perhaps 13,000 of them. It was at most about 1500. The smoke of the cannonade had drifted away and it was a bright, clear day with the temperature at 87 degrees and the humidity high. In well-dressed ranks the troops stepped off smartly behind the shelter of Seminary Ridge and began to advance at a steady 110 steps a minute, some brigades a bit slower than others.

Frank A. Haskell

A native of Vermont, Haskell (1828-1864) attended Dartmouth, graduating in 1854 as president of his class. He then moved to Wisconsin, where his brother had a law practice, and read law with a rival firm. He was admitted to the bar and practiced in Madison, gaining a reputation as a "young man of bright prospects." Meanwhile he was active in the militia and when the war broke out entered the *6th Wisconsin* as a first lieutenant. He became regimental adjutant and soon attracted the attention of Brigadier General John Gibbon, then commanding the *Iron Brigade*, of which his regiment formed a part. Haskell served with distinction as Gibbon's aide during the Second Bull Run campaign and at Antietam, Fredericksburg, Chancellorsville, and Gettysburg, where he was instrumental in steadying the line on Cemetery Ridge on 3 July and in bringing up reinforcements to break the "High Tide." After Gettysburg, Haskell was promoted to command the *36th Wisconsin*, which he had to recruit. He brought his regiment east in the spring of 1864 only to be killed leading his command in the bloody Union disaster at Cold Harbor in June of 1864. His *The Battle of Gettysburg*, a lengthy letter written to his brother after the battle, is, despite its excessive floridness, a particularly valuable source of information on the events on Cemetery Ridge on 3 July. A fine officer, Haskell never received any particular reward for his services on 3 July, despite the fact that his distinguished and courageous conduct was specifically mentioned in the report of every senior officer present on Cemetery Ridge that afternoon. Gibbon himself wrote, "I have always thought that to him, more than to any one man, are we indebted for the repulse of Lee's assault."

Union signalmen on Little Round Top spotted the movement almost as soon as it began. They flashed word of it to Cemetery Ridge, where the troops were already engaged in a myriad activities, bringing up ammunition, fetching water, carrying off the wounded. The artillerymen were resiting the guns, cleaning and reloading them. At the sight of the advancing Confederate troops, men cried out, "Here they come! Here they come! Here comes the infantry!" Even as the Rebel ranks emerged from the wooded slopes of Seminary Ridge, Union commanders bolstered their line. Hays, commanding the right end of the ridge, packed both brigades of his division into his front, intermingling them to get the

Arthur James Lyon Fremantle

Son and grandson of veterans of the Coldstream Guards, Fremantle (1835-1901) entered the British army in 1852 and was by 1860 a captain in the guards and a lieutenant colonel in the army. In 1863 he secured six months leave and paid a lengthy visit to the Confederacy, entering from Mexico, and touring much of the South, where he was made welcome. Reaching Richmond in June, Fremantle was introduced to virtually every prominent political leader in the Confederate government from Jefferson Davis on down. He soon attached himself to the Army of Northern Virginia and marched with it from 20 June to 9 July, being present at Gettysburg. Fremantle then successfully passed through the Union lines on 9 July, reaching New York on the 12th, just in time to witness the outbreak of the Draft Riots. He took

ship for England on 15 July and there wrote *Three Months in the Southern States*, which was published soon after in England and in the South.

Returning to service, Fremantle eventually rose to a lieutenant generalship and a knighthood. Although he served in the field during the 1881 campaign in Egypt, Fremantle seems never to have been under fire during his entire career save for the three days of Gettysburg. Although his book is unabashedly pro-Confederate, it is nevertheless valuable for many of the details of the battle and occasional insights which seem to have slipped past his bias. Fremantle was one of a number of European aristocrats who visited the armies during the war, many of whom actually bore arms for one side or the other.

maximum number of men on the line. Webb moved all the troops in his brigade to the stone wall along the Angle, where Alonzo Cushing, already hit twice, directed the resiting of his three remaining pieces, while with one hand he attempted to keep his intestines from spilling out of a wound in his abdomen. Gibbon, their division commander, brought up additional guns and positioned two regiments closer to the front as a ready reserve. Soon, on a thousand yards of front from Zeigler's Grove southwards, there were some 5,750 infantrymen in 27 regiments, supported by 500 artillerymen with 23 guns. Out on picket duty about 150 yards in front of the lines was the entire *8th Ohio* along the Emmitsburg Road on the right, and skirmishers strung out along the front in the open fields. The men were steady, if anxious, as they ignored

The Confederate Order of Attack

The deployment of the Union regiments along Cemetery Ridge has never been in doubt. Not so the precise order of the regiments within each of the attacking Confederate brigades, for no original documents survive which contain them. What follows is a reconstruction of the order of the regiments in the brigades which constituted the attacking column, given from left to right and front to rear.

Pettigrew's [Heth's] Division
Brockenbrough's Brigade: 55th Va.-47th Va.-40th Va.-22nd Va. Bn
Davis's Brigade: 2nd Miss.-55th N.C.-42nd Miss.-11th Miss.
Marshall's [Pettigrew's] Brigade: 11th N.C.-26th N.C.-47th N.C.-52nd N.C.
Fry's [Archer's] Brigade: 5th Ala.-7th Tenn.-14th Tenn.-13th Ala.-1st Tenn.
Trimble's [Pender's] Division
Lane's Brigade: 33rd N.C.-18th N.C.-28th N.C.-37th N.C.-7th N.C.
Scales' Brigade: 38th N.C.-13th N.C.-34th N.C.-22nd N.C.-16th N.C.
Pickett's Division
Garnett's Brigade: 56th Va.-28th Va.-19th Va.-18th Va.-8th Va
Kemper's Brigade: 3rd Va.-7th Va.-11th Va.-24th Va.
Armistead's Brigade: 38th Va.-57th Va.-53rd Va.-9th Va.-14th Va.

the occasional artillery round which still came over, and watched almost with admiration as the Rebels advanced. Union Lieutenant Colonel Edmund Rice, whose *19th Massachusetts* lay right in the path of the attack on the left of the *II Corps* front, noted that many of his men were impressed by "the grandeur of [an] attack of so many thousand men," and Gibbon's aide Haskell later wrote, "More than half a mile their front extends, more than a thousand yards the dull gray masses deploy, man touching man, rank pressing rank, and line supporting line. The red flags wave, their horsemen gallop up and down; the arms of eighteen thousand men, barrel and bayonet, gleam in the sun, a sloping forest of flashing steel. Right on they move as with one soul, in perfect order, without impediment of ditch, or wall or stream, over ridge and slope, through orchard and meadow, and cornfield, magnificent, grim, irresistible."

The attacking brigades had not only to advance under Federal artillery fire, but also had both to shift front and to

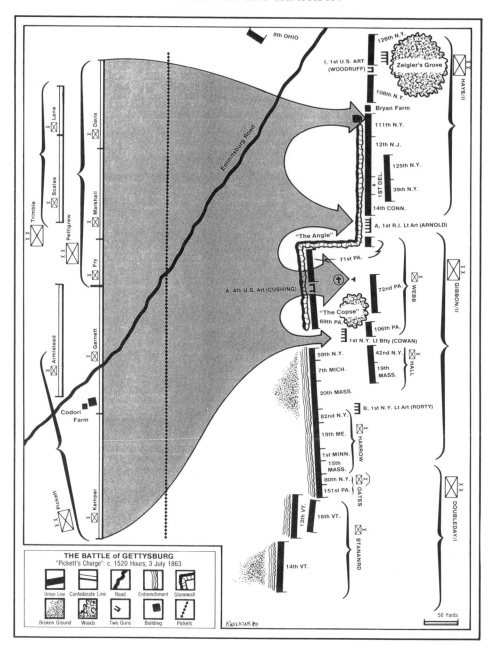

THE BATTLE of GETTYSBURG
"Pickett's Charge": c. 1520 Hours, 3 July 1863

Union Line | Confederate Line | Road | Entrenchment | Stonewall
Broken Ground | Woods | Two Guns | Building | Pickets

50 Yards

KWILKINS '85

Mrs. Hancock's Dinner Guests

On 15 June 1861 there was a farewell party held at Los Angeles for a number of officers who had resigned their commissions to "go South." It was by Captain and Mrs. Winfield Scott Hancock; the guest of honor was Colonel Albert Sidney Johnston, who died leading a Confederate army at Shiloh in early 1862. It was an unusual dinner in many ways, not least of which was in its aftermath. Years later Mrs. Hancock observed that three of the officers present died in the attack against her husband's position along Cemetery Ridge on 3 July 1863. Two of these were the then Major Lewis A. Armistead—who gave Mrs. Hancock a prayer book on that last happy occasion—and the then Captain Richard B. Garnett, both of whom led brigades against Hancock's *II Corps*. Mrs. Hancock could not recall the name of the third, nor has his identity been established with any degree of satisfaction.

close the considerable gap between them if they were to strike the Union lines with full force. Their peculiar deployment, angled obliquely at about 20 degrees to the Union lines, with Pickett's men about 325 yards in advance and to the right of Pettigrew's, had been dictated by the terrain. Seminary Ridge, behind which they formed ranks, was angled thusly. And right up against the eastern side of the ridge, in the middle of the ground over which they had to advance, was Spangler's Woods, a densely overgrown area. These had dictated the deployment, and now the troops had to engage in various maneuvers to correct their lines. As the Yankees on Cemetery Ridge looked on with amazement, Pickett's men marched directly forward for about 450 yards, and then, even as they came under artillery fire, imperturbably faced 45 degrees to their left to advance for another 250 yards. At this point their left flank halted in a depression which offered some cover, while their right advanced some 300 yards more, pivoting 30 degrees to dress ranks with the left. Meanwhile, Pettigrew's men advanced, moving about 850 yards to also halt in the same depression. By this time, roughly 1510, the two columns were no more than 300 yards apart. The shallow ground gave the troops some respite from the increasing volume of artillery fire, and they paused for a minute or two to reform their ranks and dress their lines. They were but 700 yards from Cemetery Ridge. In addition to the 23 guns which were firing into their front from Cemetery Ridge, the men were being enfiladed by 29 guns from Cemetery Hill on their left. There were 43 more guns on their right on Little Round Top and in the grand battery positioned at the lower end of the ridge, but these were as yet unable to fire effectively, for Pickett's men were partially sheltered by the berm along which ran the Emmitsburg Road. Surprisingly the troops in front took the fewest losses, for the guns on Cemetery Ridge were virtually out of long range ammunition. Those on the left flank fared less well. Subject to the attentions of the guns on Cemetery Hill, Brockenbrough's small brigade on Pettigrew's left was getting the worst of the pounding. Enfilading fire repeatedly tore through the ranks, killing and wounding many. The brigade was keeping together well, the men closing ranks to fill the gaps, but they had suffered grievously and morale was becoming brittle. By this time Rebel skirmishers were now trading shots with their Union counterparts. The advancing Rebels kept on. Pickett's men now faced somewhat leftwards and advanced another 250 yards, emerging over the berm of the Emmitsburg Road to receive the full attentions of the batteries on the Union left, while Pettigrew's men pressed on for another 125 yards. Within minutes, probably at about 1520, the columns converged with Pickett's left meeting Pettigrew's right. All through the advance Confederate officers had been keeping their men together. Longstreet, observing from the rear, had several times passed messages to Pickett and Pettigrew. Pickett decided that victory was at hand and sent back for reinforcements to follow-up the imminent breakthrough. Both he and Pettigrew stayed close to their men, riding back and forth among the lines, with a steady stream of staff officers to keep them fully informed of all developments. The injured Garnett—one of the few officers to participate in the attack mounted—rode at the head of his men. Armistead, who had placed his old black hat on the end of his sword to hold high overhead as an improvised flag found, much to the amusement of his men, that the point kept piercing the cloth so that he repeatedly had to put it back at the tip. Nevertheless, despite this light note the going had been grim. Many had been killed and wounded, and some men had taken to the rear.

Even as the two columns converged, Union infantry began to go into action. The *8th Ohio* had formed up directly beneath Cemetery Ridge, a few yards off to the left of the flank of the attacking columns. Its commander, Lieutenant Colonel Franklin Sawyer, had 200 of his own men, plus 75 stray skirmishers from the *125th New York*. He formed them into a single line and attacked right into the flank of over 12,000 battle hardened veterans. It was audacious and it worked. Sawyer's men headed right for the flank of Brockenbrough's already shaken brigade. The brigade broke even before contact was made, officers and men alike fleeing to the rear in disorder, losing three colors and many prisoners. As Pickett attempted to shift some regiments to cover the flank, Sawyer pulled his men back, placing them behind a fence from whence they began laying down a voluminous fire. The advance continued. At the point of convergence Pettigrew's men faced slightly left and Pickett's right to align the two columns. Now side-by-side, the two divisions plunged forward absorbing their skirmishers. As they covered the last 400 yards to the Union lines, the troops began to pick up the pace. The intrepid Hancock rode among the Federal skirmishers fully exposed to the enemy, who gallantly refrained from firing upon him. Union skirmishers fired off a few last rounds, forcing Garnett's brigade to halt and deliver a volley, and then scuttled for safety up the ridge. The firing from Cemetery Ridge became more intense as the artillery began delivering canister, great shotgun-like bursts containing scores of musket balls, which tore wide swathes in the ranks. With their officers exhorting them to maintain a steady pace, the Confederate infantry pressed on, now so close to the front that the batteries on Cemetery Hill and at the south end of the Ridge could no longer fire upon them. They began to receive rifle fire. Brigadier General George J. Stannard's brigade of 1,950 Vermonters from Doubleday's division was positioned

An engraving showing a portion of the Union line on Cemetery Ridge. The large tree suggests that the position of the artist was behind the lines just north of the Angle, no details being shown of the lines just south of it.

on the Union left, a bit forward of the main lines, and was the first to open fire at about 1530. Though none of the men had been in action before the previous day, they were as steady as veterans, putting out volley after volley into the front of Kemper's brigade. Unable to take the fire, the Virginians drifted to their left, narrowing the attacking front from a little more than a thousand yards to less than 600. The entire weight of the attack was directed right at the troops on either side of the conspicuous copse of trees near the Angle in the stone wall atop the ridge, right in the center of which was Webb's brigade of Philadelphians and Cushing's depleted battery.

At 250 yards every infantryman in the *II Corps* who was able to do so opened up with his rifle. At the same time, the artillerymen began firing double loads of canister. A storm of lead smashed into the face of the advancing troops. Hundreds fell. Confederate cohesion, so magnificently displayed as they had advanced for some 20 minutes under fire, dissipated. Men broke ranks, clustering together about their officers and their flags. Some fled, but most pressed on, giving a cheer and running up the slope. Davis' and Pettigrew's brigades and portions of Fry's drove for the Union right, held by Hays' men. Kemper's brigade was being forced leftwards by fire on its front and flank, delivered by Stannard's Vermonters and Gibbon's left-hand brigades under Brigadier General William Harrow and Colonel Norman J. Hall. Meanwhile, in the center Garnett's men and most of Fry's pressed forward towards Webb and the Angle. Kemper fell. Garnett was enveloped in a shell burst from which his bloodied horse alone emerged. Yet still the troops pressed on. Atop Cemetery Ridge the determined horde that began sweeping up the deceptively shallow-looking slope intimidated the faint hearted. Some men attempted to flee until their officers brought them to their senses, sometimes with bared steel.

On the extreme right of *II Corps'* line in front of Zeigler's Grove, Brigadier General Alexander Hays had intermingled several regiments from his division. Colonel Sawyer's *8th Ohio* and the stray skirmishers from the *125th New York* were still firing into the enemy flank from behind the fence in the fields, about 200 yards from Zeigler's Grove. Hays had two New York regiments directly in front of the grove, supported by Woodruff's Regular Army battery. To their left, there were some farm buildings belonging to Abram Bryan, a black man who had wisely fled with his wife and two teenaged sons when the Rebels invaded Pennsylvania. Among these and along the upper reaches of the stone wall were the *111th New York* and the *12th New Jersey*, the latter armed with old smoothbore muskets which they had loaded with buckshot. To their left were the *1st Delaware* and the *14th Connecticut*,

Brigadier General Richard B. Garnett was killed leading his brigade at Pickett's Charge. His body was never found though his sword was discovered in a pawnshop years after the battle.

with two companies carrying Sharps breechloading rifles. The left of the division rested on Captain William A. Arnold's battery, *A, 1st Rhode Island Light Artillery.* Two more New York regiments were positioned immediately behind those on the line. Gallantly led by Colonel J.K. Marshall, Pettigrew's own brigade struck, with that of Brigadier General Joseph R. Davis to its left and that of Brigadier General James Lane from the second line, yet further left, where Brockenbrough ought to have been. The 13 regiments of North Carolinians and Mississippians had been reduced to small bands huddled around their colors and their officers, but they swept up the slope bravely to receive a storm of fire full in the face. Despite

Richard B. Garnett

Born on a plantation in Virginia, Garnett (1817-1863) graduated 29th in the West Point class of 1841, joining the infantry. He served against the Seminoles, on the frontier, on staff duty during the Mexican War, and the "Mormon War," rising to captain before resigning in May of 1861 to take service with the Confederacy as a major of artillery. That November he was promoted brigadier general and given the famed Stonewall Brigade, which he led during Thomas "Stonewall" Jackson's Shenandoah Valley campaign in early 1862, during which he was charged but never tried for disobedience of orders at Kernstown. He commanded George Pickett's old brigade in the Antietam and Fredericksburg campaigns, and later in North Carolina and eastern Virginia for several months. During the Gettysburg campaign, his brigade of five Virginia regiments, was in the center of the first line of the Confederate assault force on 3 July, forming the left of Pickett's front. Having not yet fully recovered from a horse kick received some days earlier, Garnett led his brigade mounted, one of the few officers in the attack to do so. He advanced calmly to the front, ignoring shot and shell, until about 20 yards from the Federal line on Cemetery Ridge, when he and his horse were engulfed in a shell burst. The horse, Red Eye—the second finest mount in the Army of Northern Virginia, valued at $1,400—emerged mortally wounded, Garnett was seen to fall. His body was not found after the battle, probably having been buried in a mass grave after being stripped by a looter. Years later his sword turned up in a pawnshop in Baltimore. His cousin, Brigadier General Robert S. Garnett, graduated two places ahead of him in the same class at West Point, had a very distinguished career in the Old Army, resigned in 1861 to become a Confederate brigadier general, and was mortally wounded that July, becoming the first general officer of either side to die in the war.

the fact that they were taking fire in their flank from the *8th Ohio*, some of Lane's men managed to get among the farm buildings between Zeigler's Grove and the stone wall, only to be pinned and surrounded by the *111th New York*—the survivors eventually surrendering. The 26th North Carolina attacked right into the front of Arnold's battery, only to disintegrate under a wave of lead, two men alone reaching the Union lines, one still bearing the flag; holding their fire, the Yankees induced the two to surrender. Some said that before falling a private from the 11th Mississippi planted the colors on the Federal lines; others, that not a man of the regiment got within 10 feet. Though not 150 men followed Lieutenant Colonel J.A. Graves up the slope, the 47th North Carolina almost breached the lines, only to be pinned to the forward slope. Unable to advance or withdraw, they too were soon forced to surrender. The right side of the Union line had held. As the survivors of the attack attempted to retire, they were subject once more to the withering fire which they had faced as they advanced.

The left of the front along Cemetery Ridge was that held by elements of Doubleday's division of *I Corps*, with Stannard's Vermonters and a regiment each of New Yorkers and Pennsylvanians. Their fire had driven Kemper's Brigade to the left towards the Union center. Thinking fast, Stannard saw a magnificent opportunity before him. If he could bring his brigade into the fields, he might take the attackers in their flank as they smashed into the center. The imperturbable Hancock saw the opportunity as well and putting spurs to horse rushed along the rear of the lines to issue the necessary orders. Gibbon too sensed the opportunity and began running towards his left, where lay the largely unscathed brigades of Harrow and Hall. But the maneuver took time and seconds now counted.

On Gibbon's right was Webb's brigade, holding the Angle in the stone wall with two regiments—the *69th* and *71st Pennsylvania*—supported by three guns from the wounded Cushing's battery, *A, 4th Artillery*, and five of Captain Andrew Cowan's *1st Battery, New York Light Artillery*. It wasn't much, perhaps 275 rifles and 7 guns to hold more than 250 yards of front, though some 450 men from the *72nd* and *106th Pennsylvania* were just coming up. It was a bit after 1530 when, giving the famed Rebel yell for the first time in the charge, 15 regiments of the decimated but still game Virginians, Tennesseans and Alabamians of Fry, Garnett and Armistead smashed right into Webb's Philadelphia Irishmen. As the mass of Rebel infantry swarmed up the ridge, most of the *71st Pennsylvania* broke, fleeing 100 yards to the rear before being rallied by Lieutenant Haskell at sword's point with a

Alonzo H. Cushing

A native of Wisconsin, Cushing (1841-1863) graduated from West Point at 19, ranking 12th in the class of June of 1861, in which George Custer was 34th. He served in the artillery, fighting in the Bull Run, Peninsula, Antietam, and Fredericksburg campaigns. At Gettysburg he was a first lieutenant commanding *Battery A, 4th Artillery*, a Regular Army outfit. On the third day his battery was atop Cemetery Ridge, at the Angle against which the Confederate grand assault was launched. Cushing stood at his post despite the loss of three of his six guns and most of his gunners, and with two severe wounds, one in the lower abdomen. As the Confederate troops swarmed over the wall he personally discharged his last round of canister before falling over his gun with a fatal wound to the head.

Cushing was a fine, dedicated young officer. His younger brother was the naval hero William B. Cushing, who personally destroyed the Rebel ironclad CSS *Albemarle* in 1864 using a spar torpedo, and who led a storming party of sailors and marines at Fort Fisher in early 1865, dying on active duty in 1874 as a commander.

little help from Webb. As they fled, the Rebel infantry came over the top of the stone and rail fence. The gallant Cushing ran his last gun right up to the fence to fire a last round of canister into the attacking ranks and then fell dead with his third wound of the day, a bullet through the mouth which tore through the back of his head. Hat still on the tip of his sword, Armistead reached the fence. Crying, "Give them cold steel!" he led his men over it. Webb attempted to throw in the *72nd Pennsylvania*. It delivered a heavy fire but, despite the efforts of both Webb and Haskell, refused to advance. As more Confederates came over the wall, Armistead's heroic band overran Cushing's guns, fighting hand to hand with the surviving gunners under Sergeant Frederick Fuger and some heroic holdouts of the *71st Pennsylvania* under Sergeant Major John Stockton. The issue hung in the balance. As Haskell rode to bring up Hall's brigade, Webb walked calmly right across the front, passing within yards of Armistead, to get to the left of his brigade, where the *69th Pennsylvania*, having denied its right to the breakthrough, was calmly pouring fire into the enemy. He ordered it to fire without regard for friendly

Lewis A. Armistead

A North Carolinian, Armistead (1817-1863)—whose uncle, Colonel George Armistead, defended Fort McHenry against the British on the occasion of the writing of "The Star Spangled Banner"—went to West Point in 1834, but was expelled in 1836 for breaking a dinner plate over the head of Jubal Early. Commissioned directly into the Regular Army in 1839, he served in garrison and in Mexico (two brevets), rising to captain. In California at the outbreak of the war, he participated in a famous dinner at the home of Winfield Scott Hancock before resigning in May of 1861. He entered Confederate service as a major, was almost immediately promoted colonel, and given the 57th Virginia. In April of 1862 he was promoted brigadier general and given a brigade in Brigadier General George E. Pickett's division of Longstreet's command. He led his brigade bravely in every campaign of the Army of Northern Virginia from the Peninsula to Fredericksburg, served with Pickett's Division in the Carolinas in the spring of 1863, and returned to the Army of Northern Virginia in time for Gettysburg. At Gettysburg, Armistead's brigade was unengaged until the third day. During Pickett's Charge the brigade formed up behind the two leading brigades of the division, but as it advanced it gradually pressed forward, merging with the first line, so that Armistead, his hat on the end of his sword, was the first man over the wall at the Angle. As his gallant band came under increasing fire from rallying Union forces, Armistead fell mortally wounded. A good officer with a talent for leading a brigade.

troops, and brought a hundred men of the *106th Pennsylvania* to their support. This daring measure momentarily held the growing band of Confederates in the Angle. Armistead fell mortally wounded in front of an abandoned gun about 30 yards inside the Union lines. But the Rebels were in the Angle in strength. Rebel infantrymen began drifting into the copse of trees. If more came over the wall and if they could break out of the Angle, the entire front might split. More troops would be needed. The five 3-inch rifles of Captain Andrew Cowan's *1st New York Light Artillery Battery*, having with the aid of Hall's men beaten off an attack from a large but disorganized group of Kemper's men, fired into the mass from the left. Hunt rode up and, while shouting at Cowan to avoid hitting friendly troops, emptied his pistol at the enemy

Brigadier General Lewis A. Armistead fell mortally wounded after his troops managed to break the Union line for a short while before being forced back.

before his mount was hit and he went down. Hall knew what to do without awaiting orders. Even as Haskell rode up Hall wheeled three of his regiments—the *59th New York*, the *7th Michigan*, and the *20th Massachusetts*—90 degrees to their right and threw them into the copse of trees on the right of the *69th Pennsylvania*. Hancock came up behind the *19th Massachusetts* and the *42nd New York*; reining in his mount, he pointed towards the copse, shouting, "Forward, men! Forward! Now's your chance!" Colonel Rice got his Massachusetts regiment moving almost instantly with the New Yorkers following behind. Minutes later Brig. Gen William Harrow, to Hall's left, brought his three regiments and two companies of sharpshooters into the trees as well. Soon after Colonel Theodore Gates brought his *80th New York* and the *151st Pennsylvania* from Doubleday's division into the trees and beyond. A furious fight developed as the Confederates sought to break out of the Angle. Gibbon was hit, falling

The height of Pickett's Charge in which the attacking Confederate force of 13,000 men lost half its strength as casualties. This etching taken from a sketch by illustrator Alfred Waud is perhaps the only visual eyewitness account.

unconscious. Hancock's saddle was struck and pieces of wood and leather and nail were driven into his thigh. Minutes earlier the loss of these two might have had decisive res··lts, but no longer. The Confederacy's moment had passed. As the Federal regiments crowded into the copse of trees sorted themselves out, an overwhelming volume of fire swept the Angle. The Union line advanced. A Confederate gun near the Peach Orchard opened up, sending a shot tearing through the crowded Union ranks. Desperate hand-to-hand fighting ensued. Within minutes it was over as hundreds of Rebels surrendered in the Angle and before the stone wall, and yet more hundreds fled as Meade looked on, cheering and waving to his troops. The Confederate tide had crested and begun to ebb. The time was about 1540. It had taken perhaps 10 minutes from the time Armistead reached the stone wall for the Union troops to smash the attack. Yet even if Armistead's daring band—which at its peak numbered perhaps 350 men—had broken through the line, it would have meant little, for nearly 13,000 men—virtually the entire Union reserve—were converging on the spot on Meade's orders.

Even as the Confederate regiments recoiled from Cemetery Ridge, they came under new pressure. Hays threw the *8th Ohio*, the *126th New York*, and a mass of skirmishers in against the left flank of the retreating Confederates, while Stannard brought the *16th* and *13th Vermont* against their right. Pressed from behind by Gibbon's men, the retreating Rebels were now subject to a murderous cross-fire, even as they were pounded by artillery from Cemetery Ridge in their rear. The terrible battering was relieved only when the Confederate right flank guard brigades, those of Wilcox and Lang, tardily came up, supported by a resumption of artillery fire from Seminary Ridge. These diverted Stannard's attention briefly and by the time he had beaten them off, the main body of the attackers had reached the safety of the Confederate lines, where Lee came forward to meet them and offer his apologies. By about 1600 hours it was all over. The great attack by means of which the Army of Northern Virginia was to smash the *Army of the Potomac* had itself been smashed.

Armistead's Last Words

As Confederate Brigadier General Lewis A. Armistead lay mortally wounded in the middle of the Angle atop Cemetery Ridge he uttered a cry for help, adding the phrase "as the son of a widow." As the Rebel tide was ebbing by then, some men on the line received permission to go to his aid. A romantic tradition has it that one of them was First Sergeant Frederick Fuger of Cushing's battery. Fuger had known Armistead in the prewar Army, when both had served in the *6th Infantry* and in Utah. According to one of Fuger's fellow gunners, Christopher Smith, Armistead said to Fuger, "I thought it was you, sergeant. If I had known that you were in command of that battery I would never have led the charge against you." They carried Armistead to the rear and a surgeon was summoned. When he arrived, Captain Henry H. Bingham saw that he could do nothing for the dying man. He offered to see that his personal effects were sent to his family. Meanwhile, a messenger came up. He explained that he was from Major General Abner Doubleday, who had heard that a Confederate general—possibly Longstreet himself—had been captured. The messenger politely inquired as to Armistead's rank. Armistead responded by saying, "Tell General Doubleday in a few minutes I shall be where there is no rank." Bingham and Armistead engaged in some small talk, and Armistead learned that his old friend Hancock had been wounded at about the same time he had been shot. Armistead, whom Bingham described as being "seriously wounded, completely exhausted, and seemingly broken spirited" gave his watch for safekeeping to one of the onlooking officers. Then he said, "Say to General Hancock for me, that I have done him, and you all, a grievous injury, which I shall regret the longest day I live." Then Armistead was carried to a field hospital where he died on 5 July.

Although many observers have attempted to discredit or place sinis-

Altogether it had taken no more than an hour. No one kept an accurate record, but it appears that, including many hundreds of prisoners, more than 50 percent of the attacking force had become casualties, some 6,500 men, one of the highest loss rates in the war. There was but one uninjured officer above the grade of captain in the brigades of Kemper, Garnett, and Armistead. Some regiments had losses as high as 85 percent. In contrast, Union casualties had been comparatively light, totalling perhaps 1,500 men, roughly 25 percent of those engaged. Gibbon's division, which bore the brunt of

ter connotations on them, there seems no reason to doubt the authenticity of any of the words attributed to Armistead in his last hours. The phrase "as the son of a widow," is, of course, a Masonic password, but there is nothing mysterious in that, nor do we have to attribute Armistead's rescue to brother Masons, as he was obviously a high ranking officer. Given that grievously wounded men often utter strange phrases, the words are more likely to have been inspired by the pain from which Armistead was suffering at the time than any sinister conspiracy. The exchange of words with Sergeant Fuger is also probable, for by the end of the fighting in the Angle the gallant sergeant was apparently on the line with the *72nd Pennsylvania*, whose troops brought Armistead to the rear. Few have connected Armistead with Doubleday's account, yet it seems reasonable to do so, in as much as Armistead was the only Confederate general taken that day. The greatest controversy has been inspired by Bingham's account, for to some the words attributed to Ar-

mistead suggested that in the hour of his death a Confederate general had denied the righteousness of the Southern cause. Northerners made much of this, sometimes embroidering upon it to make it sound stronger. Southerners, on the other hand, reacted to the statement by denying that Armistead had said any such thing. Despite this, Dr. Bingham's account has a ring of truth to it and was reported rather casually. Indeed, after the war, when his veracity was under attack from unreconstructed Rebels, he wrote out a careful statement in which he described the circumstances and events in great detail, even noting that he was uncertain as to whether Armistead had said "regret" or "repent." This ought to have stilled all criticism, but the matter had passed beyond reason and become part of the myth of the "Lost Cause." These words and their clumsy construction ought to be taken for what they were, the painful utterances of a dying man who had no personal animosities towards those of his friends against whom he found himself fighting.

---

the attack, lost about 40 percent, but no other major unit lost more than about 18 percent. In the ultimate contemporary measure of victory, Union troops had captured at least 28 Confederate battle flags—possibly as many as 33—without the loss of a single one: Colonel Arthur F. Devereux of the *19th Massachusetts* was found to have 4 Confederate colors draped over his arm when the smoke cleared. Firing continued for a while, but as the afternoon wore on the battlefield fell silent once more.

The Union lines shortly after the repulse of Pickett's Charge. The position is apparently just to the south of the Angle, as shown by the copse of trees on the left of the view.

As surgeons offered what help they could to the wounded, the troops savored the results of the day's action. On the Union side there was great elation. It had been a hard fought battle, but at the moment of crisis everyone had done his duty. The troops themselves had the most praise for Hancock, who was mobbed by cheering men even as he lay in a hospital bed—game as ever, he dictated a report to Meade in which he recommended an immediate counterattack, and then attempted to address the troops before passing out. But it was the *Army of the Potomac* itself that was the victor. A fine, skilled, professional force, Gettysburg was the first battle in which it was ably led on all levels and permitted to fight to the end. Meade, neither as brilliant a strategist nor as able an administrator or inspiring a leader as either McClellan or Hooker, was unlike either of them willing to fight.

The troops had been greatly disappointed with the results of Chancellorsville. When Meade gave them their chance, they took it. Almost everyone had performed well, occasionally magnificently. Only one incident marred the day's success. Even as the great charge was being met, Brigadier General Judson Kilpatrick, commanding the *3rd Cavalry Divi-*

Where Was Pickett?

After the failure of the attack on Cemetery Ridge, Major General George Pickett came in for considerable criticism concerning his conduct during it. Many of the recriminations revolved around his location at various critical moments during the attack, most notably at the time of the final rush up the ridge. A number of hostile accounts suggested, and many flatly stated, that he deliberately avoided the post of danger, preferring to hang back rather than lead his men into the mouths of the Yankee artillery. This is hardly fair, since it was the task of brigadier generals to lead the men forward, as was ably done by Garnett, Kemper, and particularly Armistead. The post dictated by duty for a major general was in the rear, directing the advance, and shifting brigades about as needed. This Pickett did well. He kept the advance going, issued the appropriate orders for dressing the lines, and changing the direction of march. He also maintained contact with Longstreet and his brigadiers by a steady stream of messengers. During the final moments of the attack he was at the Codori farm, about 200 yards from Cemetery Ridge, about in the center of where his division was supposed to be, although Kemper's brigade had slipped northwards. The only fault—and that a grievous one— which can be attributed to the general, came after the climax of the attack. At the time Armistead's daring band was swarming over the stone wall into the Angle, Pickett was standing alone near the Codori house, having dispatched all four of his aides on various missions. As the Union troops on the fringes of the area under attack streamed down the ridge to take the attackers in the flanks, he appears to have decided that the game was up. He remounted his horse and rode slowly off to the rear, leaving his men to fend for themselves in their retirement.

How different the conduct of Brigadier General James J. Pettigrew! Pettigrew, too, properly remained in the rear of his division as it advanced, keeping in contact with Longstreet and his brigadiers. But when the assault faltered he remained on the field, directing the retirement, trying to get a rear guard organized. Despite a wound in the hand he stayed with the troops until the end, being among the last to regain the security of Seminary Ridge. It was against Pettigrew that Pickett's performance was measured and found wanting, though to blame him for the failure of the attack is unreasonable, for his personal failure came after the climax of the action.

How Many Men Fought at Gettysburg?

Estimates of the number of men engaged at Gettysburg vary widely. Figures for the strength of the Army of Northern Virginia range from a low of about 65,000 men of all arms to a high of about 75,000, while those for the *Army of the Potomac* vary between 80,000 and 95,000. On paper the total strength of the two armies was enormous: the Army of Northern Virginia officially had over 100,000 men in early June and the *Army of the Potomac* nearly 200,000. But in practical terms, that is men physically present; the Army of Northern Virginia apparently numbered about 77,000 men at the beginning of June and the *Army of the Potomac* about 115,000—plus about 13,000 cavalry horses, 7,000 artillery horses, and 9,000 horses and 22,000 mules to haul its 5,200 supply wagons and ambulances, roughly one for every 22 men. These are the strengths with which the armies began to march northwards into Pennsylva-

nia. But these are not useful figures. The rolls of both armies and particularly the *Army of the Potomac* counted great numbers of non-combatants. There were, in fact, several different ways which the troops could be counted. In the Union service these may be summarized as:

Present: including all personnel for whom rations had to be issued.

Present for Duty: excluding personnel on sick call or under arrest, but including teamsters, hospital personnel, musicians, and other non-combatants.

Present for Duty Equipped: including only combat-ready enlisted personnel, and their officers, the Confederates did not use this category, referring instead to "Effectives," a figure which applied to enlisted men only, to the exclusion of officers, who could number up to 7 percent additional personnel.

Normally, accounts of Civil War battles rely for the numbers of troops rather heavily on the Union

sion on picket duty covering the Union left flank, had been instructed to harass the exposed Confederate right flank. Through the afternoon, his troopers had a modest degree of success, forcing Hood to put several regiments into line to cover his flank. At about 1730, Kilpatrick learned of the successful defense of Cemetery Ridge. He decided to attack. Overruling the objections of one of the most brilliant young officers in the army, Brigadier General Elon J. Farnsworth— whom he virtually accused of cowardice—Kilpatrick ordered a charge directly into the ranks of Hood's infantry. The result was predictable. Though nearly 100 prisoners were taken,

"Present for Duty Equipped" figures and a modified version of the Confederate "Effectives." However, this is not necessarily accurate, particularly as such figures are almost never available for the specific date of a battle. In the case of the Gettysburg campaign, Union "Present for Duty" figures were reported on 30 June for most formations, while comparable Confederate figures are available for different units over the previous few days. Between the reporting dates and the battle itself, many men became stragglers, others were sent off on detached duty, and some men took sick, while in some instances, additional men and units reported for duty. Upon including all of these changes, a new set of figures results, "Engaged" strength, personnel actually committed to combat, whether they fought or not.

Engaged Strengths of the Armies at the Battle of Gettysburg

Day	Union (Loss)	Confederate (Loss)	US:CS (Loss)
1 July	23.4 (12.0)	42.7 (8.0)	1:1.8 (1:0.7)
2 July	71.6 (8.0)	49.9 (8.0)	1:0.7 (1:1.0)
3 July	63.6 (3.0)	53.9 (12.0)	1:0.8 (1:4.0)
4 July	60.0 (-)	41.9 (-)	1:0.7 (-:-)
Total	93.0 (23.0)	69.9 (28.0)	1:0.8 (1:1.2)

This table compares the forces actually on the field during each day of the battle of Gettysburg, including the cavalry action of 3 July. All figures are in thousands, and for each day take into account the forces which were on hand at the start of the day's fighting and those which arrived during it, with losses in parentheses. Losses are deducted from each day's total before the balance is carried forward to the next day. The US:CS figures give the ratio of Federal figures to those of the Confederacy.

about 20 percent of the 300 troopers from the *1st Vermont Cavalry* fell, and Farnsworth himself was killed, taking five wounds. Considerable compensation for the disaster was the fact that Brigadier General David Gregg's *2nd Cavalry Division* had successfully clashed with Stuart's cavalry several miles east of the main action on the Rummel farm, between the York Pike and the Hanover Road, beating them off and once again demonstrating that Southern superiority in the mounted arm was a thing of the past. Thus, Kilpatrick's foolhardy venture did not mar the general elation in the Union ranks. When Meade appeared on Cemetery Ridge within minutes of the breaking of the attack, the troops cheered him.

There was no cheering in the Confederate ranks, only great dejection. Despite the skill, determination, and courage of its men, the Army of Northern Virginia had failed in its most severe test, one which it had undertaken with the greatest confidence. And that perhaps more than any other reason was why it had failed. A sympathetic visiting British officer, Colonel Sir Arthur J. L. Fremantle of the Coldstream Guards, would note in his diary, "It is impossible to avoid seeing that the cause of this check to the Confederates lies in the utter contempt felt for the enemy by all ranks." Colonel William C. Oates of the 15th Alabama put it even more succinctly when he wrote of Lee, "he was overconfident." The failure was a terrible blow to Lee. At one point he cried out, in an agonized tone, "Too bad! *Too bad!* **Oh! Too bad!**" Now, with his grand attack broken and more than two divisions in fragments, he had to face whatever Meade might have in store.

As it turned out, Meade had nothing in store though he considered making a counterattack. Even after beating off the Rebel attack Meade had available plenty of ammunition: over a dozen fresh batteries and lots of relatively fresh troops—the virtually unscathed *VI Corps* plus the bulk of *V Corps*, the

Cavalry commander Elon J. Farnsworth. He attained the rank of brigadier general shortly before the battle of Gettysburg where he would be killed leading an ill-fated charge on 3 July.

A dubious depiction of Farnsworth's charge. It is interesting for its depiction of the horses, which are not shown in typical pre-photography gait.

battered but rested *III Corps* and parts of both *I Corps* and *XII Corps*. But he rightfully thought better of it. Even his uncommitted forces were tired after days of marching under the hot sun and the army's rations had practically run out. Nor was the tactical situation suitable. His best available fresh outfits, the *VI* and *V Corps*, were over on the left in the area of the Round Tops. To shift these troops into the center would take time and expose them to a flank attack from Longstreet's Corps. Moreover, the Rebel batteries still lined Seminary Ridge and, although they were low on shot and shell as a result of the bombardment, he knew very well that they still had plenty of canister. He essayed a probe of the Confederate right flank, only to find that Longstreet's troops were still full of fight. In addition an attack there would be over the same rugged ground on which the previous day's fighting had been. Either option seemed ill-advised. The only way Lee could still win the battle was for the *Army of the Potomac* to risk an offensive. Time was on Meade's side. If he waited he could expect reinforcements and supplies within a few days, while Lee's situation would deteriorate further. The better part of wisdom was to wait and see what Lee had in mind.

Who Won the Third Day?

The final day's fighting was clearly a Union victory. For this, Lee must ultimately bear a large share of the responsibility, perhaps all of it. Despite the heroic way in which the troops conducted the day's attack, it had been an unwise, possibly even a foregone conclusion. The battle had been lost on the second day, when, despite horrendous losses, the *Army of the Potomac* had attained the security of Cemetery Ridge. Lee ought to have heeded Longstreet's advice and avoided a direct clash entirely. To have attempted a frontal assault against superior forces was at best foolhardy, given the numerous demonstrations of the superiority of the defense over the offense which had occurred throughout the entire war. Barely six months earlier the *Army of the Potomac* had itself essayed a frontal assault at Fredericksburg only to be repulsed at great cost with but trivial losses to the Confederacy. On 3 July Meade held a somewhat less favorable position than Lee had held at Fredericksburg, but with substantially more men. The best course of action for 3 July would have been to pull Ewell's Corps back over on the left and stand on the defense along Seminary Ridge, tempting Meade into an attack. Had Meade made such

an effort, the result could well have been a Union debacle. And if he failed to attack, or tried to engage in some fancy maneuvers, Lee could pull his army back across a series of readily defensible ridge-lines until he could break off contact and retire.

In the *Army of the Potomac* it seems as if everyone did well. Meade, who after all had only been in command five days, kept his head, lightly supervising his subordinates, all of whom were doing their jobs properly, and moved his reserves as he saw the need develop. The true hero of the day was undoubtedly Hancock, who was constantly at the post of danger and did exactly what had to be done to break Lee's grand attack. Alpheus Williams had done nicely in beating off Ewell's blow on Culp's Hill earlier in the day, though Slocum's ill-advised interference had caused one serious reverse. The other corps commanders had all done their part in essentially support roles. Hays, Gibbon, and Doubleday had handled their divisions well, certainly contributing to the successful outcome of the fight. Several brigade commanders—Webb, Hall, Harrow, Stannard, Greene, and Kane in particular—had performed with great resourcefulness and courage.

CHAPTER VIII

The Retreat

4 - 26 July

*T*here were intermittent rains all through the morning of 4 July, cooling and refreshing the tired troops of both armies. By then Lee had pulled back Ewell's Corps and brought it into line alongside Hill's, making a continuous front about 7,000 yards long from Oak Hill southwards along Seminary Ridge. The last of the ammunition was served out and it was discovered that the army had enough for one more day of heavy fighting. Hoping Meade would attempt an attack, Lee had his men dig rifle pits and trenches on the western side—the reverse slope—of Seminary Ridge, under cover of the woods. Meanwhile he laid plans for a retreat. That morning he sent his supply train with its precious cargo of wounded men and booty off under escort. One column, 17 miles long, went northwestwards through Cashtown, while a shorter one went to the southwest through Fairfield. Lee held his troops in the trenches all day, hoping Meade would strike.

Meade refused to attack, disregarding even Lincoln's polite urgings that he do so. In the circumstances this was wise, though he might perhaps have thrust a strong column towards the mountains in an effort to canalize Lee's retreat. Instead, Meade confined himself to some probes with cavalry and infantry, while the army rested. Anticipating needs over the next few days, he instructed Major General Darius Couch, commanding the emergency volunteers and militia along the

A famous photograph of a fallen Confederate sharpshooter in Devil's Den that turned out to be faked. Many of the pictures of the battle of Gettysburg were posed by photographers who dragged bodies around to set up a dramatic scene. The body here was also photographed in a different location.

Susquehanna and Major General William H. French, who had 11,000 good troops available at Frederick and Maryland Heights, to be prepared to advance on short notice. Meanwhile, he arranged for rations to be brought up for the hungry troops, some of whom had not eaten in two days, and began bringing out the wounded, many of whom were found in houses and barns and cellars in Gettysburg now that the Rebels had pulled back from the town. Thousands of stragglers were returned to their units by the provost marshal. Ammunition was inventoried and distributed, with Hunt

The shallow grave of soldiers killed at Gettysburg. Some 5,000 troops died at Gettysburg and many of these were hastily or improperly buried; others were not interred for some time. A collection by the Union states who lost soldiers at the battle provided for a military cemetery at Cemetery Hill. Confederate troops were buried in mass graves, but were reinterred in Richmond after the war.

calculating that the average artillery piece had fired 100 rounds so far during the battle and that there remained unexpended over 100 more per gun.

By afternoon, with the rains coming down more heavily, Lee knew he would have to move. It would be a painful retreat, for nearly 7,000 seriously wounded men would have to be left behind, but he had no choice if he was to save the rest of the army, now reduced to not more than 45,000 effective troops. In the evening A.P. Hill's corps set out for Fairfield, followed shortly afterwards by Longstreet's; with Ewell's stepping off at about 0200 on 5 July. The cavalry was out covering his flanks and the men marched with grim determination.

During the night Meade received evidence that a retreat was underway. He arranged for Gouverneur Warren to

Two Confederate soldiers wish a wounded comrade farewell as Lee dispatches his wounded to the rear.

conduct a probe with Major General John Sedgwick's *VI Corps*. Meanwhile he issued orders for the army to move southwards. Warren encountered Ewell's rear guards near Fairfield at about 1500. A small skirmish resulted, little more than an exchange of artillery fire at long range, but enough to convince both Warren and Sedgwick that Lee had deployed his entire army in Fairfield Gap. Meade immediately ordered the army to march to the support of Sedgwick. On Meade's instructions, Sedgwick essayed another probe on 6 July, which provoked a spirited response. Meade, who had briefly toyed with the idea of pursuing Lee through Fairfield Gap, decided that such an advance would be unprofitable if the gap was held in strength. In addition, he had received information to the effect that Lee was closing on Hagerstown, further south. Telling a couple of brigades to follow Lee's rearguards through the gap as they withdrew, Meade put the *Army of the Potomac* on the road for Frederick. Meanwhile, Union cavalry harassed the retreating Confederate supply trains, inflicting some losses. Particularly hot little actions occurred near Greencastle, at Hagerstown, and near Williamsport, as Buford's and Kilpatrick's troopers, seeking to impede the Rebel retreat, traded blows with Stuart's over the next few days.

It was not until the morning of 7 July that Meade tardily began a really determined pursuit of Lee, with some units pressing on as much as 34 miles despite extremely heavy

The General Atmosphere

Accounts of Gettysburg rarely mention the strong odor which must have hung over the field. The troops on both sides had been marching for weeks in seasonably hot weather with little opportunity to bathe. As a case in point, the 15th Alabama appears to have had its first bath in weeks sometime after 8 July, when the Army of Northern Virginia reached the security of the old Antietam battlefield; it did not have another until 23 July. Add to the stench of tens of thousands of unwashed human bodies that of tens of thousands of horses, plus three days accumulated excreta from both species along with the progressive decay of thousands of dead men, and animals, and by 4 July the location of the armies must have been readily discernable from some distance downwind.

rains. By 9 July the *Army of the Potomac* was drawn up on a front roughly five miles long running southwards from Boonesborough to Rohrsville, just east of the old Antietam battlefield. The army had successfully shifted its base to Frederick, less than 15 miles to the rear and in direct rail communications with Washington and Baltimore. There a great supply dump was being set up. Many men needed shoes after the exhausting march and there was a shortage of fresh horses for the cavalry and artillery. Even as the army was moving into this position, Meade effected some organizational changes. The higher officer ranks of the army had been badly depleted in the battle with Reynolds dead and Hancock, Sickles, Gibbon, Birney, and Butterfield wounded. Major General John Newton now had Reynolds' *I Corps*, Major General William H. French would take Sickles' *III Corps* from Birney and Brigadier General William Hays would take over Hancock's *II Corps*. Brigadier General Alexander Webb took over Gibbon's division. Butterfield's loss was by no means unwelcome. Relieved on 5 July, his position as chief of staff had been cooperatively and ably exercised temporarily by Major General Alfred Pleasonton, the chief of cavalry, and Brigadier General Gouverneur K. Warren, the chief of engineers. Meade now made Brigadier General A.A. Humphreys his chief of staff, giving his division to newly arrived Briga-

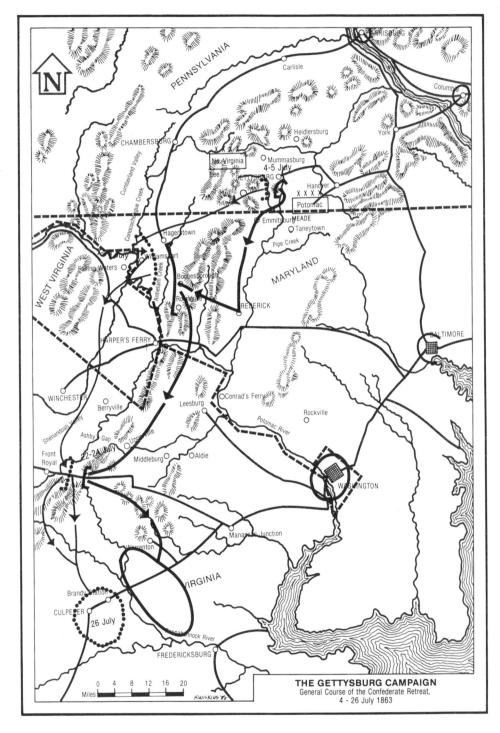

THE GETTYSBURG CAMPAIGN
General Course of the Confederate Retreat,
4 - 26 July 1863

dier General Henry Prince. All of the new men were compe-
tent, but none compared with Reynolds and Hancock, whose
loss Meade felt deeply, both as personal friends and as
brilliant commanders. Another loss of great concern to Meade
was the many thousands of veteran troops who had been
killed, wounded or captured. Moreover, due to peculiarities
of Federal recruiting policies, he now also faced the loss of
additional thousands of men through expiration of their
enlistments, including all of Stannard's fine brigade of Vermonters and several other regiments as well. Fortunately,
replacements had begun to come in. The return of stragglers
had brought the army up to about 65,000 men. Now 8,500
more from French's old command in Western Maryland were
incorporated bodily, mostly into *III Corps*, with a few going to
II Corps. He also received about 1,500 cavalry returning to
duty after being remounted, 6 fresh batteries and a single new
long-service regiment, the *39th Massachusetts*. An additional
10,500 short service men were forwarded from Baltimore and
XVIII Corps on the Virginia-Carolina coasts. This was a net
gain of about 21,500 men, but only about half of them were
reliable enough, or had sufficient enlistment time left, to
permit their incorporation into the army. There were other
troops on hand as well, totalling over 30,000, but these were
all emergency short-service men enlisted during the invasion
of Pennsylvania. Meade ordered them to advance cautiously,
to pose a potential threat to Lee's army, and to restrict his
foraging, but knew that they were completely overmatched
by Lee's veterans. Certain political leaders, including Lincoln, failed to comprehend this essential point. To all appearances Meade, with over 100,000 men, was hesitating in face of
an enemy with little more than half his numbers. Pressed by
Washington, Meade won General in Chief Henry Halleck to
his views. By now Meade was convinced that Lee intended to
make a stand north of the Potomac, which was in flood,
concentrating his army for battle west of the old Antietam
battlefield. Intending to mass his entire strength before seeking a battle, Meade moved with extreme—perhaps excessive—caution from the Boonesborough-Rohrsville line,

taking three days to advance but eight miles. Lee had gotten his entire army into the Hagerstown area by 7 July only to abandon the town five days later to establish himself on a broad front of about a dozen miles from Hagerstown southwards along Antietam Creek towards the Potomac. There Meade was in position to seek battle with Lee who, of course, had already made extensive preparations to meet him.

Lee had little inclination to effect an immediate crossing of the Potomac, which was in flood in any case, and set about preparing for battle. By 11 July he had established a sound defensive line on a seven-mile front curving from Conococheague Creek on his left to the Potomac, on his right. This put water lines in his rear, making it theoretically an unfavorable position, but the terrain was good and there were two normally good crossing points in his rear at Falling Waters and Williamsport. He had limited communications with the other bank, permitting him to transfer his wounded and bring up some badly needed supplies, and his engineers were working on repairing a pontoon bridge damaged by Union raiders on 6 July. In addition, he fortified the entire front and prepared a second line behind it to cover the two towns. Aside from the cavalry, he tried to keep his troops out of contact with the enemy. Stuart's cavalry was kept busy, however, covering the flanks and keeping Union cavalry away from the main body. Skirmishes continued to occur on an almost daily basis. These were mostly small affairs, but there were occasional larger ones, notably near Hagerstown and at Funkstown. By 13 July, however, Lee's position was sufficiently secure to permit him to pull in his cavalry as well, and it assumed a position on the flanks of the main line. Then he waited, hoping that Meade would attack him.

Meade intended to attack on 13 July, and so informed Halleck. But he was still new at his job, having been in command only a fortnight. Conducting a successful offensive battle was a different matter from conducting a successful defense. He convened a conference of his senior officers at about 2000 on 12 July. There he proposed a reconnaissance in force which could be converted into a proper attack should the opportunity present itself. Five of the corps commanders (Hays, *II*; French, *III*; Sykes, *V*; Sedgwick, *VI*; and Slocum, *XII*) expressed total opposition to the plan, with only two favoring it (Wadsworth, *I*, and Howard, *XI*), along with Humphreys, Warren, and Pleasonton. As the bulk of his field commanders, including the two who were technically senior to him, Slocum and Sedgwick, opposed the plan, Meade felt it would be unwise to make the attempt. This went against his own judgment, for he had perhaps 75,000 men, albeit many were green, and Lee had no more than 55,000, though veterans all. He postponed the attack until he could make a personal survey of the Confederate lines. He spent 13 July doing just that in pouring rain, accompanied only by Humphreys, against a background of heavy skirmishing between the two armies. Upon its completion, he issued orders substantially implementing the plan for 0700 on the morning of 14 July. By then it was too late, for Lee had flown.

Lee had hoped that the *Army of the Potomac* would attack on 13 July. When it failed to do so and when the Federal troops began entrenching themselves, he remarked, rather unfairly in view of the reception he had prepared for them, "They have but little courage!" He ordered a retreat. By this time his engineers had completed repairs to the Falling Waters bridge and the Potomac had gone down to the point where it could be forded in places. The movement started just after dark, with Longstreet and Hill's men using the bridge and Ewell's the ford at Williamsport. Despite the heavy rains, the move was well executed. By noon on 14 July only two divisions, Heth's and Lane's (formerly Pender's), remained north of the river. Scouts sent out by Brigadier General Horatio G. Wright (*1st Division/VI Corps*) got wind of the movement early on 14 July. Wright immediately sent his skirmishers forward. Discovering that the enemy had evacuated their trenches, he pushed his entire division towards Williamsport. Sedgwick followed with the rest of the corps, and at 0830 Meade ordered a general advance. The cavalry went in first with Kilpatrick's division of about 3,500 sabers pushing up to Falling Waters. There he surprised Heth and some 3,000

The trains of the **Army** *of the Potomac at Berlin, on the Potomac River, about two weeks after the battle. Often overlooked in accounts of the war is the enormous logistical effort which went into supporting armies in the field.*

infantrymen entrenched on a ridge about 2,500 yards from the town. Kilpatrick made a hasty, ill-conceived attack with but two squadrons. Though Pettigrew, actually commanding the Confederate rear guard, received a mortal wound during the attack, it was a complete failure. Both squadrons were almost eliminated as combat formations. As Kilpatrick renewed his attack in more formal style, Buford's division came up and hit Heth's troops in their right rear. The Rebels put up a tough rear-guard fight, finally getting across the river in about an hour with a loss of some 700 prisoners and a few wagons. The Army of Northern Virginia had crossed the Potomac and reached the security of the Shenandoah Valley.

Both armies now began to move eastwards, Lee trying to return to his former lines south of the Rappahannock and Meade attempting to intercept him. The *Army of the Potomac* crossed the river whose name it bore near Harper's Ferry on 17 July and began moving down the eastern side of the Blue Ridge Mountains, roughly paralleling Lee's line of march in the Shenandoah Valley about 15 to 20 miles to the west. Both armies had cavalry out, and several small skirmishes took place. On 21 July there were larger clashes at Manassas Gap and Chester Gap. These convinced Meade that Lee's army was in the vicinity. On 22 July he ordered Major General William French to force Manassas Gap with his *III Corps* in preparation for a general attack. Unfortunately, French proceeded in a rather lethargic fashion, allowing Confederate light forces to block his advance. As a result his corps did not get through the gap until the next day, only to find Lee deployed before Front Royal with two corps—Longstreet's and Hill's—in line and his third—Ewell's—nearby. French halted immediately. Meade began to get the balance of the *Army of the Potomac* forward, anticipating a general attack for 24 July. That night, Lee slipped away again.

Manassas Gap was the last important engagement of the Gettysburg campaign. On 24 July Ewell's Corps began arriving at Culpeper Courthouse, just south of the Rappahannock River. Hill and Longstreet came up soon after. On 26 July the *Army of the Potomac* began encamping not five miles away, just north of the Rappahannock. Despite nearly two months of maneuver and battle the two armies were virtually back where they started from. The Gettysburg campaign was over.

Military Leadership in the Gettysburg Campaign

During the Gettysburg campaign two of the most skillful armies in the world maneuvered and fought against each other for nearly two months. One of them, the Army of Northern Virginia, had had the better of the other, the *Army of the Potomac*, in virtually every encounter for well over a year. Yet in this campaign, the Army of Northern Virginia suffered a devastating defeat. One reason for this was the leadership of each army. The senior officers on both sides were the products of the same military educational system, virtually everyone being a West Pointer. All had imbibed the concepts and principles of Napoleonic warfare as expounded by Henri Jomini, with its rationalistic systems. It is unlikely that any had ever heard of Karl von Clausewitz, let alone read any of his books on military science.

Hitherto the military leadership of the Confederacy had been generally superior to that of the Union. There were many reasons for this. The South was essentially an aristocratic society, drawing its senior officers primarily from the landed planter class, men who were bred to command and men whom less privileged Southerners were quite willing to follow. Moreover, at the start of the war some very fine officers of the Old Army took service with the Confederacy. Many of these men embodied both the high technical competence of the professional soldier and the aristocratic character of the planter class. In addition, the Confederacy lacked a regular army. As a result, all of the approximately 300 Regular U.S. Army officers who joined the Confederate Army quickly attained high rank in the newly created Provisional Army, so that a great deal of natural and professional leadership was immediately available to command the field armies.

In contrast, the North was essentially a democratic society, with a deep suspicion of the Regular officer, who was widely viewed as an aberrant character in a democracy. Moreover, there was a widespread belief that war was not a technically difficult profession. In addition, the Union retained the services of the Regular Army. This army already had a number of senior officers, and it was these men who led the Union forces in the initial stages of the war. However, the qualities necessary to attain high rank in peacetime are not usually those needed in war. Furthermore, although there were many good men among the 767 active officers who remained loyal, most of them were left with the Regular troops, rather than being spread among the volunteers. One result was that the younger officers who first reached high rank were those who had left the Army before the war and reentered it as volunteers in 1861, such as George McClellan, William

Tecumseh Sherman, Henry Halleck, Ambrose Burnside, Joseph Hooker, and Ulysses S. Grant.

As the South was on the strategic defensive, its generals enjoyed a considerable advantage over those of the Union, for the changing technology of war had come to greatly favor the defensive. The result was that the initial batch of Union commanders was not particularly successful, while the Confederate commanders gained a considerable reputation for competence. Of course, as time went on and defeats accumulated, better officers rose to command the armies of the Republic. This was critically important in terms of the eventual outcome of the war, for Union leadership grew steadily better during the struggle, while Confederate leadership, good at the start, did not improve markedly. The brilliance of Robert E. Lee and several of his subordinates generally obscures the fact that most Confederate commanders were rather ordinary, deriving their reputations as much from the inherent advantage of the defensive as from any particular talent for war. By mid-1863 the differences between the military leadership on both sides had begun to disappear. Certainly there was little to choose from between Union or Confederate company and regimental officers, and probably not much difference between their brigade commanders. But the Confederacy still retained something of an advantage in the higher levels, at least in the Army of Northern Virginia.

Robert E. Lee was certainly one of the finest defensive strategists in history, and a resourceful, imaginative, and daring tactical commander. James Longstreet was one of the best tacticians in American history, being meticulous, thorough, and brave, though with no talent for independent command. Ambrose P. Hill, and Richard Ewell had proven able division commanders, but during the Gettysburg campaign were each directing a corps for the first time. This was a matter which ought to have caused Lee more concern than seems to have been the case, particularly as Ewell had an independent streak, much preferring to be left alone to do things in his own way and on his own terms. The South had its political generals too. Virginian A.P. Hill had been chosen by fellow-Virginian Lee to command Third Corps over the more capabl⌐ South Carolinian Daniel H. Hill. Indeed, Virginians were noticeably over-represented in upper ranks of Lee's army: only 1 of 3 lieutenant generals and 4 of 11 major generals were not Virginians, while Virginia supplied but 7.5 of the 37 infantry brigades and 5 of the 7 cavalry brigades involved in the campaign, most of which were commanded by men from their native states.

Among Confederate division commanders, two of Longstreet's subordinates, Lafayette McLaws, and John B. Hood, were particularly able. Jubal Early was perhaps less skillful than either of these but was certainly more aggressive, although rather insubordinate. The rest were all seasoned officers of proven ability. J.E.B. Stuart was, of course, a fine cavalryman, particularly capable when needed to screen movements or to secure information, but he had an unfortunate thirst for glory. Lower ranking Confederate commanders were all generally good, some of them particularly so.

The leadership of the *Army of the Potomac* was not in the same class as that of the Army of Northern Virginia. Joe Hooker had been, and would subsequently again prove to be, a good corps commander. As an army commander he was an able administrator and a particularly brilliant strategist, but he lacked the intestinal fortitude and imagination to lead an army successfully. His replacement, George G. Meade, was neither as brilliant a strategist nor as able an administrator or an inspiring leader as Hooker. He was, however, willing to fight. Cautious and unimaginative he might have been, but he possessed a determination and dogged pugnacity which his predecessors lacked. Moreover, he had the confidence of his subordinates. Among the Union corps commanders, both John Reynolds and Winfield Scott Hancock were excellent, both being resourceful, bold, and aggressive men, fully the equal of Hill or Ewell, and certainly of Longstreet. The others were capable, if unspectacular. Dan Sickles, the last political general left with the army, had little professional skill but was commendably aggressive in a command which often lacked aggressive generals. Alfred Pleasonton was by no means as brilliant a cavalryman as Stuart nor indeed as his own subordinate John Buford, but he was reliable, understood his job, and did what he was supposed to do. Most of the Union division commanders were good, and several were particularly so, including James S. Wadsworth, Abner Doubleday, John Gibbon, Andrew A. Humphreys, Alpheus Williams, and John Newton, a number of whom eventually rose to command corps. Lower ranking Union commanders were by this time quite skilled.

All of this leads to the conclusion that the leadership of the Army of Northern Virginia was still considerably better than that enjoyed by the *Army of the Potomac*. Yet in the end, it was the Army of Northern Virginia which suffered defeat. There were a number of reasons for this. Lack of a well-defined Confederate strategy was certainly a factor. Beyond this is the fact that during the Gettysburg campaign, Lee and his generals were performing rather poorly while their opponents were doing rather well.

In the end, battles go to the side which makes the fewest mistakes. At Gettysburg, for several reasons, that was the Union. As a result the battle assumed a significant importance, for Lee's defeat at Gettysburg was not merely a tactical or strategic reverse. It was also a moral disaster. The *Army of the Potomac* had long been superior to the Army of Northern Virginia in numbers and equipment and had proven itself equal to its foe in technical skill and courage. And now it had proven equal in leadership.

CHAPTER IX

The Gettysburg Campaign and the War for the Union

*A*lthough the Civil War was essentially won in the Western Theater, the Gettysburg campaign was the central event of the great conflict. These seemingly contradictory statements are both true. To seek victory in the Western Theater implied a long, hard struggle. Such a conflict was one for which neither Union nor Confederate political and military leadership was prepared nor understood. Both sides thought in terms of a short war with immediate decisive results, harking back to Napoleon's many swift victories or the war with Mexico, rather than to the more accurate model of the lengthy American Revolution. For the North, a protracted war meant unprecedented mobilization and repeated offensives to wear down the Confederacy's resistance. For the South, such a war meant a repetitive series of defensive battles designed to erode the Union's will. This was highly unpalatable to both sides. A quick win was what both were seeking, and a quick win could only be gained in the Eastern Theater. But a quick win proved elusive and by mid-1863 some perceptive individuals were beginning to understand that the war would be long. It was against this background that the Gettysburg campaign ran its course. The greatest offensive effort of Confederate arms, the campaign represented perhaps the only opportunity that the South had to win the war by offensive means. The Confederate loss at Gettysburg meant that the war would go on. The South would never again have

Gettysburg, 19 November 1863, showing the parade which preceded Lincoln's famous address.

so fine an army as that which marched north from Culpeper Courthouse in June of 1863, while the armies of the Republic would continue to grow stronger and better. In a protracted war, the greater material resources of the Union would inevitably triumph.

July of 1863 saw three great defeats inflicted upon the South, for Union success in the East at Gettysburg was matched in the West by the surrender of Vicksburg to Major General Ulysses S. Grant on 4 July, and the completion of Major General William S. Rosecrans' Tullahoma operation (26 June-4 July), which secured central Tennessee for the Union. Together these three events mark the central moment of the war in both a literal and a figurative sense. In July of 1863 the war had run nearly two years since Bull Run and had nearly two years yet to go before Appomattox. Gettysburg, Vicksburg, and Tullahoma were signals that the tide of battle had turned definitively against the South. Thus, in a very real sense, the Gettysburg campaign was the "High Tide" of the Confederacy.

Lincoln at Gettysburg

On the cool afternoon of Thursday 19 November 1863, a little more than four months after the battle of Gettysburg, President Abraham Lincoln delivered what is perhaps the most well known speech in American History, a model of the oratorical art. The occasion was the dedication of a portion of Cemetery Hill as a national cemetery. Although the distinguished politician and orator Edward Everett was to be the principal speaker of the day, Lincoln had been asked to say a few words.

Lincoln wrote the first draft of the address at the White House, not on a train. And he did not write it on the back of an envelope. He began it on a piece of White House stationery. As originally composed, it had but eight sentences. He subsequently added a ninth in pencil which ran over on to another piece of paper. This was the version he brought with him to Gettysburg on 18 November.

That night, both Lincoln and Everett were guests at the home of David Wills, a local Republican Party official. At some point during the night Lincoln revised his speech, writing a second draft on two fresh sheets of paper, and adding the final tenth sentence.

The ceremonies occupied much of the next day, which was rather overcast. The 70-year-old Everett spoke for two hours. When he was finished a patriotic ode was pro-nounced by a large chorus. Then Lincoln was introduced. Lincoln held the speech in his left hand, referring to it only occasionally, as he spoke slowly in his somewhat nasal, mid-western voice.

"Four score and seven years ago our fathers brought forth, upon this continent, a new nation, conceived in Liberty, and dedicated to the proposition that all men are created equal.

"Now we are engaged in a great civil war, testing whether that nation, or any nation, so conceived, and so dedicated, can long endure. We are met here on a great battlefield of that war. We have come to dedicate a portion of it as a final resting place for those who here gave their lives that that nation might live. It is altogether fitting and proper that we should do this.

"But in a larger sense we can not dedicate—we can not consecrate—we can not hallow—this ground. The brave men, living and dead, who struggled here, have consecrated it, far above our poor power to add or detract. The world will little note, nor long remember, what we say here, but can never forget what they did here. It is for us, the living, rather to be dedicated here to the unfinished work which they have, thus far, so nobly carried on. It is rather for us to be here dedicated to the great task remaining before us—that from these honored dead we take increased devotion to that cause for which they gave the last full measure of devotion—that we here highly resolve that these dead shall not have died in vain; that this nation shall have a new birth of freedom; and that this government of the people, by the people, and for the people, shall not perish from the earth."

Contrary to tradition, applause interrupted Lincoln five times during his brief address. At the end the crowd gave what one observer called "tremendous applause" and three cheers. Despite Lincoln's belief that the address was a "flat failure," its oratorical merit was recognized immediately. Everett put it best when he told Lincoln, "...there was more in your two minutes than in my two hours." History and popular sentiment have proven him right.

Guide for the Interested Reader

*T*his is not intended to be a comprehensive bibliography of the Gettysburg campaign, but rather a guide to further inquiry for those with a more than passing interest in the campaign and the Civil War.

Recommended Reading

The Civil War is perhaps the most widely-written-about subject in American history. There has been a steady stream of serious works for almost 130 years. This listing on the Gettysburg campaign is by no means to be considered definitive. Rather, it is a short guide for further reading. The serious student of Gettysburg should consult Richard A. Savers' *The Gettysburg Campaign, June 3-August 1, 1863: A Comprehensive, Selectively Annotated Bibliography* (Westport, Ct: 1982) for a more significant guide to the literature of the battle and the campaign.

The single most valuable work on Gettysburg is undoubtedly Edwin B. Coddington's *The Gettysburg Campaign: A Study in Command* (New York: 1968/Dayton, Oh: 1979), which is exhaustively researched and meticulously detailed. All other works are essentially useful supplements to this, or expand upon particular aspects of the campaign in much greater detail. Also of great value is *The United States Army War College Guide to the Battle of Gettysburg* (Carlisle, Pa.: 1986).

Of particular value are Edward J. Stackpole, *They Met at Gettysburg* (3rd Ed., Harrisburg, PA: 1982), a classic account which remains of considerable value; Glenn Tucker, *High Tide at Gettysburg* (Indianapolis: 1958) and *Lee and Longstreet at Gettysburg* (Indianapolis: 1968), both careful accounts of the battle, done from an essentially Southern point of view, and the second is notable for its discussion, and refutation, of the criticisms levelled at Longstreet as well as for an examination of other problems during the campaign; Wilber S. Nye's *Here Come the Rebels* (Baton Rouge: 1965) is a very good account of the operations of the Army of Northern Virginia in Pennsylvania, with considerable detail on many of the small skirmishes which took place prior to Gettysburg; John W. Schildt, *Road to Gettysburg* (Pasner, WV: 1978) covers the same events with more detail but is much less readble; David G. Martin's *Gettysburg: July 1* (Conshohocken, PA: 1995) is a very detailed treatment of the events of 1 July, while Henry W. Pfanz' *Gettysburg: The Second Day* (Chapel Hill: 1987) and *Gettysburg—Culp Hill and Cemetery Hill* (Chapel Hill: 1994) provide an extraordinary look at the events of 2 July; Oliver W. Norton's *The Attack and Defense of Little Round Top* (New York: 1913) is a very detailed treatment of that critical moment in the battle; George R. Steward's *Pickett's Charge* (2nd Edition, Dayton, OH: 1980) is a meticulous account of the circumstances and events of the Confederate attack on Cemetery Ridge on 3 July, done in great detail; Fairfax Downey's *The Guns at Gettysburg* (New York: 1958) is a good account of the artillery during the battle, with considerable detail on the lives and work of artillerymen and of their service, while his *Clash of Cavalry* (New York: 1959) is the best account in English of Brandy Station—the best work on the subject is in German—and does an excellent job of illustrating the life and work of cavalrymen, their mounts, and their service; the role of the cavalry in the campaign is best treated in Edward G. Longacre's *The Cavalry at Gettysburg* (Rutherford, NJ: 1986), albeit that it tends to be very favorable to the Confederate perspective. A remarkable guide to the appearance of the battlefield may be found in William A. Frassanito's *Gettysburg: A Journey in Time*

(New York: 1975), which meticulously examines contemporary photographs and drawings, and which may be usefully supplemented by Vol. III of *The Image of War*, edited by William C. Davis (New York: 1983).

The First Day at Gettysburg (Kent, OH: 1993), *The Second Day at Gettysburg* (Kent, OH: 1993), and *The Third Day at Gettysburg and Beyond* (Chapel Hill: 1994), all edited by Gary Gallagher, contain valuable essays on military leadership during the battle. Biographies of Lee abound, most notably Douglas Southall Freeman's *R.E. Lee: A Biography* (New York: 1934-1935). On Longstreet see William Piston's *Lee's Tarnished Lieutenant* (Athens, GA: 1987) and Jeffry D. Wert's *General James Longstreet: The Confederacy's Most Controversial Soldier, A Biography* (New York: 1993), two very good works. Other biographies of value are Freeman Cleaves' *Meade of Gettysburg* (Norman, OK: 1960), Edward J. Nichols' *Toward Gettysburg: A Biography of General John F. Reynolds* (University Park, PA: 1958), Edgecumb Pinchon's *Dan Sickles* (Garden City, NY: 1945), Glenn Tucker's *Hancock the Superb* (New York: 1960), Manly Wade Wellman's *Giant in Gray: A Biography of Wade Hampton of South Carolina* (New York: 1949), Martin Schenck's *Up Came Hill* (Harrisburg, PA: 1958), Burke Davis's *Jeb Stuart, The Last Cavalier* (New York: 1957); Alice Rains Trulock's *In the Hands of Providence: Joshua L. Chamberlain and the American Civil War* (Chapel Hill: 1992); Edward G. Longacre's *Pickett: Leader of the Charge* (Shippensburg, PA: 1995) and *General John Buford: A Military Biography* (Conshohocken, PA: 1995); and Kent Masterson Brown's *Cushing of Gettysburg* (Lexington, KY: 1993).

There are many firsthand accounts of the battle available. The most readily accessible are those in Vol. III of *Battles and Leaders of the Civil War*, edited by Robert Underwood Johnson and Clarence Clough Buell, which has seen several editions. *Gettysburg*, edited by Early Schenk Myers and Richard Brown (New Brunswick, NJ: 1948) usefully collects a great many accounts of different incidents during the battle from a range of individuals, which includes everyone from generals down to schoolboys. Frank A. Haskell's *The Battle of Gettysburg* and

Sir Arthur Fremantle's *Three Months in the Southern States* have both also been reissued with some frequency. The memoirs of many of the participants are available, among the most interesting is Longstreet's *From Manassas to Appomattox*, which has seen several editions. Rarely available, but recently reissued, is William C. Oates' *The War Between the Union and the Confederacy and a History of the 15th Alabama* (Dayton, OH: 1980).

No student of the campaign and battle of Gettysburg can ignore the valuable work of John W. Busey and David G. Martin, *Regimental Strengths and Losses at Gettysburg* (Hightstown, NJ: 1986), on which the present work relies rather heavily.

A useful guide to the military practice of the times may be found in Albert A. Nofi's *A Civil War Treasury* (Conshohocken, PA: 1992). H.C.B. Rogers' *Confederate and Federals at War* (New York: 1983) is a useful supplement to this as is Jack Coggins' *Arms and Equipment of the Civil War* (New York: 1962), which covers in detail an enormous range from uniforms to tactics with profuse illustrations. Aside from a great deal of racial clap-trap, *Attack and Die* by Grady McWhiney and Perry Jamieson (Alabama: 1982) gives a good look at the problems of Civil War tactics. Other works of value on this subject are B.I. Wiley's two volumes, *The Life of Johnny Reb* (Indianapolis: 1953) and *The Life of Billy Yank* (Indianapolis: 1952). Mark M. Boatner III's *The Civil War Dictionary* (2nd Edition, New York: 1979) and *The Historical Times Illustrated Encyclopedia of the Civil War*, edited by Patricia L. Faust (New York: 1986), are handy for sorting out information on the conduct of war during the Civil War, on the individuals involved and on the Gettysburg campaign in particular. Also of value as reference works are the two volumes by Ezra J. Warner, *Generals in Blue* (Baton Rouge: 1984) and *Generals in Gray* (Baton Rouge: 1983).

The subject of Lincoln at Gettysburg has been explored in Philip B. Kunhardt's *A New Birth of Freedom* (Boston: 1983) and more recently in Gary Wills' *Lincoln at Gettysburg: The Words that Changed America* (New York: 1992).

There are a number of novels which deal with Gettysburg. The most recent of any value are John Jakes' *The Blue and the Gray* and Michael Shaara's *Killer Angels*.

Simulations and Wargames

There are a very large number of wargames—both board and computer—dealing with the Gettysburg Campaign in one way or another. Three classics, albeit somewhat hard to find, are:

Gettysburg (Avalon Hill, 1964)
Lee Moves North (Simulations Publications, 1973)
Terrible Swift Sword (Simulations Publications, 1976)

Among more recently published items, special note should be taken of:

Battleground Gettysburg (Talonsoft, 1996)
Dixie: Gettysburg, 1863 (Columbia Games, 1996)
Fateful Lightning (XTR, 1995)
They Met at Gettysburg (Spearhead, 1996)
Three Days of Gettysburg (GMT, 1995)
Thunder at the Crossroads (The Gamer, 1994)

Films

While the Civil War has been a central theme in a number of notable motion pictures, Gettysburg has been rather neglected. The two most significant efforts to portray the battle span a period of over 75 years. D.W. Griffith, whose father was colonel of the Confederate 1st Kentucky Cavalry, included the battle in his masterful and racist *Birth of a Nation*. The effort is, however, unrealistic and overly romanticized. More recently, the television version of John Jakes' *The Blue and the Gray* did a more ambitious job of Gettysburg, which, while inaccurate in numerous ways—including having Lincoln speak after Gettysburg but before Vicksburg—is useful for the technical details, most notably in terms of the artillery. *Killer Angels* is quite accurate in technical details, and does a wonderful job of depicting some of the most critical episodes of the battle.

The Gettysburg National Military Park

The movement to preserve the Gettysburg battlefield as a memorial park was started by the veterans themselves. After many years, their efforts resulted in the acquisition of the battlefields and adjacent land by the National Park Service. Today park land encompasses a significant portion of the original battlefield. Albeit rather crowded with monuments, the Gettysburg battlefield is surprisingly well preserved. Aside from the areas adjacent to the town of Gettysburg itself, and along some of the routes leading northwards from it, development has been relatively limited, particularly in the areas of Longstreet's attack on 2 July and Pickett's Charge on the 3rd. There has been no significant construction or erosion on the main battlefield, save in the area immediately northwest of Cemetery Hill. There has, however, been a rather considerable expansion of wooded areas so that sites which were sparsely treed in the summer of 1863 are now virtually forested. It is thus no longer possible to see clearly the area of Pickett's Charge from Little Round Top, since a grove obscures part of the view.

The field is liberally sprinkled with memorials, markers, and guns dedicated to particular units, individuals, and states. The Pennsylvania monument is the largest, but that to Brigadier General Gouverneur K. Warren on Little Round Top is perhaps the most impressive of the scores of monuments. A monument to Longstreet is planned. Many of the markers are properly sited and provide useful guides to the location of specific regiments, batteries and individuals during the battle. This is particularly the case for the markers along Cemetery Ridge. In other cases, monuments have been placed for aesthetic reasons and are of little use as guides to the battle. A number of the monuments are meticulously preserved and regularly decorated, notably that to the New York *Irish Brigade*. Others are sadly neglected. The descendants of some of the men who fought at the battle preserve their memory and it is moving to note that on both the Pennsylvania memorial and the New York memorial on Little Round Top, the names of individual soldiers have from time to time been carefully polished by their progeny.

About 25 years after the battle several hundred cannon were placed on the field as part of the preservation program. While the exact number varies, in as much as pieces are sometimes removed for repair, there are about 225 guns to mark Union positions and about 180 to mark Confederate. However, only one gun—a 3-inch rifle at the foot of Buford's statue—has been positively identified as having been at the battle. Many clearly were not, for they bear 1864 foundry marks, and a number of the pieces are facsimiles, such as 6-pounders which have been drilled out to look like 12-pounder Napoleons. The most carefully sited pieces, which are in some instances accompanied by caissons, are those along Cemetery Ridge, where one may even find one with a burst barrel, an occasional hazard of artillery service.

The Visitor Center and the Cyclorama Center provide useful exhibits on the battle, though with occasional inaccuracies and with a generally bland narrative. The cyclorama itself is rather spectacular, but not always informative. It is possible to tour the field by bus or car, but walking is undoubtedly the best way for anyone seriously interested in understanding the course of events. The field can be covered on foot in two full days, devoting one to the areas of the 1 July fighting, Cemetery Hill and Culp's Hill, and the second to the area from the Peach Orchard to Little Round Top and to Cemetery Ridge and the ground before it, saving the Angle for last.

Order of Battle

The Gettysburg Campaign
1 - 3 July 1863

This order of battle gives the regiments and batteries which comprised the *Army of the Potomac* and the Army of Northern Virginia at some time during the battle of Gettysburg, with some formations which formed part of these armies but were not present at the battle.

Abbreviations: Standard state abbreviations have been used. Unless otherwise noted a unit is always understood to be a regiment of infantry. Infantry regiments of both sides, and Confederate cavalry regiments, normally had companies A through K, omitting J; Union cavalry regiments normally had A through M, omitting J; but some regiments of both types had higher designated companies than the norm. Other abbreviations used are: **Att.**, attached; **Art.**, artillery; **Bde.**, brigade; **Btty.**, battery; **Bn.**, battalion; **Coy.**, company; **Div.**, division; **Elms.**, elements, an unspecified portion of a unit; **Gd.**, guard; **HQ**, headquarters; **Lt.**, light; **pdr.**, pounder; **Reg.**, Regular Army; **Res.**, reserves; **Vol.**, volunteer army.

Commanders. Where more than one officer is listed, the second named officer replaced the first in command during the battle, and so forth.

Symbols: The symbol * indicates that only a portion of the unit was present; **w** shows that the officer in question was wounded, **k** that he was killed, **mw** that he was mortally wounded, and **c** that he was captured; **/** separating two unit designations indicates they were serving as a single merged unit; **#** indicates that the unit was not present for the battle.

Numbers: The figures given here represent the number of men actually engaged at Gettysburg on 1-3 July, and is sometimes at variance with that given in the text, where the total strength of units has usually been indicated, which frequently includes elements not actually present. Figures for units not present are not given. As a rule, for higher formations engaged strength was about 85 percent of the total strength carried on the rolls of the individual units in the last few days of June, the difference being due to straggling, illness, and the detachment of men for special details. Numbers have sometimes been estimated. Summary figures are given for corps, divisions, and brigades and include personnel assigned to headquarters and staff duty, and may also reflect cross-assignment of units indicated in footnotes but not listed with the particular organization. If no figure is given for a unit, it was not present on the battlefield though in the general vicinity.

The Army of the Potomac: 360 guns/93,540

Commanding General: Maj. Gen. George G. Meade
Chief of Staff: Maj. Gen. Daniel Butterfield (w)
Chief of Engineers: Brig. Gen. Gouverneur K. Warren (w)
Chief of Artillery: Brig. Gen. Henry J. Hunt (w)
Provost Marshal General: Brig. Gen. Marsena R. Patrick
Assistant Adjutant General: Brig. Gen. Seth Williams
Chief Quartermaster: Brig. Gen. Rufus Ingalls
Medical Director: Dr. Johnathan Letterman
Chief of Ordnance: Lt. John R. Edie (acting)

Provost Gd., **1,365**
  93rd N.Y., 371
  2nd Pa. Cav., 489
  8th U.S. Cav., A-G & I, 401
  Casuals from *1st, 2nd, 5th,* & *6th U.S. Cav.,* 18

Guards and Orderlies:
  Oneida N.Y. Cav. Co., 42

Engineer Bde. **#**
  Brig. Gen. Henry W. Benham
  15th N.Y. Eng., A-C
  50th N.Y. Eng.
  U.S. Eng. Bn.

Signal Corps, 36
  Capt. Lemuel P. Norton

I Corps, 28 guns/12,222

Maj. Gen. John F. Reynolds (k)
Maj. Gen. Abner Doubleday
Maj. Gen. John Newton
Escort, *1st Maine Cav.*, L, 57

1st Div., 3,857
Brig. Gen. James S. Wadsworth

1st ["Iron"] Bde., 1,829
Brig. Gen. Solomon Meredith
Col. William W. Robinson
19th Ind., 308
24th Mich., 496
2nd Wisc., 302
6th Wisc., 344
7th Wisc., 364

2nd Bde., 2,017
Brig. Gen. Lysander Cutler
7th Ind., 434
76th N.Y., 375
84th N.Y. [14th Militia], 318
95th N.Y., 241
147th N.Y., 380
56th Pa., A-D & F-K, 252

2nd Div., 2,997
Brig. Gen. John C. Robinson

1st Bde., 1,537
Brig. Gen. Gabriel R. Paul (w)
Col. Samuel H. Leonard (w)
Col. Adrian R. Root (w, c)
Col. Richard Coulter (w)
Col. Peter Lyle
Col. Richard Coulter (w)
16th Maine, 298
13th Mass., 284
94th N.Y., 411
104th N.Y., 286
107th Pa., 255

2nd Bde., 1,452
Brig. Gen. Henry Baxter
12th Mass., 261
83rd N.Y. [9th Militia], 199
97th N.Y., 236
11th Pa. [1st Bde. 1 July], 270
88th Pa., 274

90th Pa., 208

3rd Div., 4,701
Maj. Gen. Abner Doubleday
Brig. Gen. Thomas A. Rowley
Maj. Gen. Abner Doubleday
Provost Gd., *Co. D, 149th Pa.*, 60

1st Bde., 1,361
Brig. Gen. Thomas A. Rowley
Col. Chapman Biddle (w)
Brig. Gen. Thomas A. Rowley
80th ["Ulster Guard"] N.Y. [20th Militia], 287
121st Pa., 363
142nd Pa., 336
151st Pa., 467

2nd ["Pennsylvania Bucktail"] Bde., 1,317
Col. Roy Stone (w)
Col. Langhorne Wister (w)
Col. Edmund L. Dana
143rd Pa., 465
149th Pa. A-C, E-J, 450
150th Pa., A-I, 400

3rd ["Paper Collar"] Bde. [Arrived 2 July], 1,950
Brig. Gen. George J. Stannard (w)
Col. Francis V. Randall
12th Vt. #
13th Vt., 636
14th Vt., 647
15th Vt. #
16th Vt., 661

I Corps Art. Bde., 28 guns/596
Col. Charles S. Wainwright
2nd Maine Lt. Btty [six 3"-rifles], 117
5th Maine Lt. Btty [six 12-pdr. Napoleons], 119
E/L, 1st N.Y. Lt. Art. [six 3"-rifles], 124
B, 1st Pa. Lt. Art. [four 3"-rifles], 106
B, 4th U.S. Art. [six 12-pdr. Napoleons], 124

II Corps, 28 guns/11,347

Maj. Gen. Winfield Scott Hancock (w)
Brig. Gen. John Gibbon (w)
Brig. Gen. William Hays
Escorts, 164
  6th N.Y. Cav., D, K, 64
  17th Pa. Cav., E, H, 100*
Provost Gd: 1st Minn., L, 48

1st Div., 3,320
Brig. Gen. John C. Caldwell
Provost Gd., 116th Pa., B, 32;
  53rd Pa., A-B, K, 70

1st Bde., 853
Col. Edward E. Cross (k)
Col. Col. H. Boyd McKeen
5th N.H., 179
61st N.Y., 104
81st Pa., 175
148th Pa., 392
6th Pa. Cav., E & I, 81

2nd ["Irish"] Bde., 532
Col. Patrick Kelly
28th Mass., 224
63rd N.Y., A-B, 75
69th N.Y., A-B, 75
88th N.Y., A-B, 90
116th Pa., A, C-D, 98

3rd Bde., 975
Brig. Gen. Samuel K. Zook (k)
Lt. Col. John Fraser
52nd N.Y. & Elms. 7th N.Y. [15], 134
57th N.Y., 175
66th N.Y., 147
140th Pa., 514

4th Bde., 851
Col. John R. Brooke (w)
27th Conn., A-B, 75
2nd Del., 234
64th N.Y., 204
53rd Pa. C-I, 135
145th Pa., 202

2nd Div., 3,608
Brig. Gen. John Gibbon (w)
Brig. Gen. William Harrow

1st Bde., 1,366
Brig. Gen. William Harrow
Col. Francis Heath
19th Maine, 439
15th Mass., 239
1st Minn., A-K & 1st Coy. Minn.
  Sharpshooters, 330
82nd N.Y. [2nd Militia], 355

2nd ["Philadelphia"] Bde., 1,224
Brig. Gen. Alexander Webb (w)
69th Pa., 284
71st Pa., 261
72nd Pa., 380
106th Pa., 280

3rd Bde., 922
Col. Norman Hall
19th Mass., 163
20th Mass., 243
7th Mich., 165
42nd ["Tammany"] N.Y., 197
59th N.Y., 152
Att.: 1st Coy., Mass. Sharpshooters, 42

3rd Div., 3,644
Brig. Gen. Alexander Hays
Provost Gd., 10th N.Y. Bn., 82

1st Bde., 941
Col. Samuel S. Carroll
14th Ind., 191
4th Ohio, 299
8th Ohio, 209
7th W. Va., 235

2nd Bde., 1,105
Col. Thomas A. Smyth (w)
Lt. Col. Francis C. Pierce
Provost Gd., 36
14th Conn., 172
1st Del., 251
12th N.J., 444
108th N.Y., 200

3rd Bde., 1,508
Col. George L. Willard (k)
Col. Eliakim Sherrill (k)

Lt. Col. James M. Bull

39th N.Y. ["Garibaldi Guard"], A-D, 269

111th N.Y., 390

125th N.Y., 392

126th N.Y., 455

II Corps Art. Bde., 28 guns/605

Capt. John C. Hazard

B, 1st N.Y. Lt. Art./14th N.Y. Btty [four 10-pdr. Parrotts], 117

B, 1st R.I. Lt. Art. [six 3"-rifles], 117

B, 1st R.I. Lt. Art. [six 12-pdr. Napoleons], 129

I, 1st U.S. Art. [six 12-pdr. Napoleons], 112

A, 4th U.S. Art. [six 3"-rifles], 126

III Corps, 30 guns/10,675

Maj. Gen. Daniel E. Sickles (w)

Maj. Gen. David D. Birney (w)

Escort, *6th N.Y. Cav., A,* 51

1st Div., 5,095

Maj. Gen. David G. Birney (w)

Brig. Gen. J.H. Hobart Ward

1st Bde., 1,516

Brig. Gen. Charles K. Graham (w, c)

Col. Andrew Tippin

57th Pa. A-C, E-F, H-K, 207

63rd Pa., 246

68th Pa., 320

105th Pa., 274

114th Pa., 259

141st Pa., 209

2nd Bde., 2,188

Brig. Gen. J. H. Hobart Ward

Col. Hiram Berdan

20th Ind., 401

3rd Maine, 210

4th Maine, 287

86th N.Y., 287

124th N.Y., 238

99th Pa., 277

1st U.S. Sharpshooters, 313

2nd U.S. Sharpshooters, A-I, 169

3rd Bde., 1,387

Col. P. Regis De Trobriand

17th Maine, 350

3th Mich., 237

5th Mich., 216

40th N.Y., 431

110th Pa., A-C, E, H-I, 152

2nd Div., 4,924

Brig. Gen. Andrew A. Humphreys

1st Bde., 1,718

Brig. Gen. Joseph B. Carr

1st Mass., 321

11th Mass., 286

16th Mass., 245

12th N.H., 224

11th N.J., 275

26th Pa., 365

84th Pa. #

2nd Bde., 1,837

Col. William R. Brewster

70th N.Y., 288

71st N.Y., 243

72nd N.Y., 305

73rd N.Y., 349

74th N.Y., 266

120th N.Y., 383

3rd Bde., 1,365

Col. George C. Burling

2nd N.H., 354

5th N.J., 206

6th N.J., 207

7th N.J., 275

8th N.J., 170

115th Pa., A-G, I-K, 151

III Corps Art. Bde., 30 guns/596

Capt. George E. Randolph (w)

Capt. A. Judson Clark

2nd N.J. Lt. Btty [six 10-pdr. Parrotts], 131

D, 1st N.Y. Lt. Art. [six 12-pdr. Napoleons], 116

4th N.Y. Lt. Btty [six 10-pdr. Parrotts], 126
E, 1st R.I. Lt. Art. [six 12-pdr. Napoleons], 108

K, 4th U.S. Art. [six 12-pdr. Napoleons], 113

V Corps, 26 guns/10,907

Maj. Gen. George Sykes
Escort, *17th Pa Cav., D & H, 78**
Provost Gd., *12th N.Y., D, E,* 99

1st Div., 3,417
Brig. Gen. James Barnes
1st Bde., 655
Col. William S. Tilton
18th Mass., 139
22nd Mass. & 2nd Coy. Mass. Sharpshooters, 137
1st Mich., 145
118th Pa., 233
2nd Bde., 1,422
Col. Jacob B. Sweitzer
9th Mass., 411
32nd Mass., A-B, D-K, 242
4th Mich., 342
62nd Pa., A-M, 426
3rd Bde., 1,336
Col. Strong Vincent (mw)
Col. James C. Rice
20th Maine, 386
16th Mich. & Brady's Coy., Mich. Sharpshooters, 263
44th N.Y., 391
83 Pa., 295

2nd Div., 4,021
Brig. Gen. Romeyn B. Ayres
1st Bde., 1,574
Col. Hannibal Day
3rd U.S., B-C, F-G, I-K, 300
4th U.S., C, F, H, K, 173
6th U.S., D-I, 196
12th U.S., A-D, G, 1st Bn., A, C-D, 2nd Bn., 413
14th U.S., A-G, 1st Bn., F-G, 2nd Bn., 490
2nd Bde., 958
Col. Sidney Burbank
2nd U.S., B-C, F, H-K, 201

7th U.S., A-B, E, I, 116
10th U.S., D, G-H, 93
11th U.S., B-G, 1st Bn., 286
17th U.S., A, C-D, G-H, 1st Bn., A-B, 2nd Bn., 260
3rd Bde., 1,484
Brig. Gen. Stephen H. Weed (k)
Col. Kenner Garrard
140th N.Y., 447
146th N.Y., 454
91st Pa., 210
155th Pa., 424

3rd Div. [Arrived 28 June], **2,853**
Brig. Gen. Samuel W. Crawford
1st Bde., 1,243
Col. William McCandless,
1st Pa. Res. [*30th Pa.*], 377
2nd Pa. Res. [*31st Pa.*], 232
6th Pa. Res. [*35th Pa.*], 323
13th Pa. Res. [*42nd Pa. "Bucktails"*], 297
3rd Bde., 1,605
Col. Joseph W. Fisher
5th Pa. Res. [*34th Pa.*], 284
9th Pa. Res. [*38th Pa.*], 320
10th Pa. Res. [*39th Pa.*], 401
11th Pa. Res. [*40th Pa.*], 327
12th Pa. Res. [*41st Pa.*], *A-I,* 272

V Corps Art. Bde., 26 guns/432

Capt. Augustus P. Martin
3rd Btty, Mass Lt. Art. [six 12-pdr. Napoleons], 115
C, 1st N.Y. Lt. Art. [four 3"-rifles], 62
L, 1st Ohio Lt. Art. [six 12-pdr. Napoleons], 113
D, 5th U.S. Art. [six 10-pdr. Parrotts], 68

I, 5th U.S. Art. [four 3"-rifles], 71

VI Corps, 46 guns/13,577

Maj. Gen. John Sedgwick
Escort, 1st N.J. Cav., L, 38
Provost Gd., 1st Pa. Cav., H, 54*

1st Div., 4,207
Brig. Gen. Horatio Wright,
Provost Gd., 4th N.J., A, C, H, 80
1st Bde.,1,319
Brig. Gen. Alfred T.A. Torbert,
1st N.J., 253
2nd N.J., 357
3rd N.J., 281
15th N.J., 410
2nd Bde., 1,322
Brig. Gen. Joseph J. Bartlett,
5th Maine, 293
121st N.Y., 409
95th Pa., 308
96th Pa., 308
3rd Bde., 1480
Brig. Gen. David A. Russell
Brig. Gen. Joseph J. Bartlett (3
June)
6th Maine, 377
49th Pa., 275
119th Pa., 403
5th Wisc., 419

2nd Div., 3,603
Brig. Gen. Albion P. Howe
2nd Bde., 1,827
Col. Lewis A. Grant
2nd Vt., 443
3rd Vt., 364
4th Vt., 380
5th Vt., 294
6th Vt., 330
3rd Bde., 1,773
Brig. Gen. Thomas H. Neill
7th Maine, B-D, F, I-K, 216
43rd N.Y., 370
49th N.Y. & Elms. 33rd New York
[60 men], 418
77th N.Y., 367
61st Pa., 286

3rd Div. 4,731
Maj. Gen. John Newton
Brig. Gen. Frank Wheaton
1st Bde., 1766
Brig. Gen. Alexander Shaler
65th N.Y., 276
67th N.Y., 349
122nd N.Y., 395
23rd Pa., 466
82nd Pa., 277
2nd Bde., 1,591
Col. Henry L. Eustis
7th Mass., 319
10th Mass., 360
37th Mass., 564
2nd R.I., 347
3rd Bde., 1,368
Brig. Gen. Frank Wheaton
Col. David J. Nevin
62nd N.Y., 237
93rd Pa., 234
98th Pa., 351
102nd Pa., 103#
139th Pa., 442

VI Corps Art. Bde., 46 guns/937
Col. Charles H. Tompkins
1st Btty, Mass. Lt. Art. [six 12-pdr. Napoleons], 135
1st Btty, N.Y. Lt. Art. [six 3"-rifles], 103
3rd Btty, N.Y. Lt. Art. [six 10-pdr. Parrotts], 111
C, 1st R.I. Lt. Art. [six 3"-rifles], 116
G, 1st R.I. Lt. Art. [six 10-pdr. Parrotts], 126
D, 2nd U.S. Art. [four 12-pdr. Napoleons], 126
G, 2nd U.S. Art. [six 12-pdr. Napoleons], 101
F, 5th U.S. Art. [six 10-pdr. Parrotts], 116

XI Corps, 26 guns/9,054

Maj. Gen. Oliver O. Howard
Maj. Gen. Carl Schurz
Maj. Gen. Oliver O. Howard
Escort, *1st Ind. Cav., I-K*, 50
Provost Gd., *17th Pa. Cav., K*, 36
HQ Gd., *8th N.Y. Independent Coy*, 40

1st Div., 2,459
Brig. Gen. Francis Barlow (w)
Brig. Gen. Adelbert Ames
1st Bde., 1,118
Col. Leopold von Gilsa
41st N.Y., A-E, G-K, 218
54th N.Y., 183
68th N.Y., 226
153rd Pa., 487
2nd Bde., 1,337
Brig. Gen. Adelbert Ames
Col. Andrew L. Harris
17th Conn., 386
25th Ohio, 220
75th Ohio, 269
107th Ohio, 458

2nd Div. , 2,775
Brig. Gen. Adolph von Steinwehr
1st Bde., 1,156
Col. Charles R. Coster
134th N.Y., 400
154th N.Y., 190
27th Pa., A-E, G-K, 277
73rd Pa., 284
2nd Bde., 1,614
Col. Orlando Smith
33rd Mass., 481
136th N.Y., 473
55th Ohio, 321

73rd Ohio, 338

3rd Div., 3,079
Maj. Gen. Carl Schurz
Brig. Gen. Alexander Schimmelfennig
Maj. Gen. Carl Schurz
1st Bde., 1,670
Brig. Gen. Alexander Schimmelfennig
Col. George von Amsberg
Brig. Gen. Alexander Schimmelfennig
82nd Ill., 310
45th N.Y., 375
157th N.Y., 409
61st Ohio, 247
74th Pa., A-B, D-K, 326
2nd Bde., 1,403
Col. Wladimir Krzyzanowski
58th N.Y., 193
119th N.Y., 257
82nd Ohio, 312
75th Pa., A-I, 208
26th Wisc., 435

XI Corps Art. Bde., 26 guns/604
Maj. Thomas W. Osborn
13th Btty, N.Y. Lt. Art. [four 3"-rifles], 110
I, 1st N.Y. Lt. Art. [six 3"-rifles], 141
I, 1st Ohio Lt. Art. [six 12-pdr. Napoleons], 127
K, 1st Ohio Lt. Art. [four 12-pdr. Napoleons], 110
G, 4th U.S. Art. [six 12-pdr. Napoleons], 115

XII Corps, 20 guns/9,788

Maj. Gen. Henry W. Slocum
Brig. Gen. Alpheus S. Williams
Provost Gd., *10th Maine, A-B, D*, 169

1st Div., 5,256
Brig. Gen. Alpheus S. Williams
Brig. Gen. Thomas H. Ruger

1st Bde., 1,835
Col. Archibald L. McDougall
5th Conn., 221
20th Conn., 321
3rd Md. [Arrived 2 July], 290
123rd N.Y., 495
145th N.Y., 245
46th Pa., 262
2nd Bde., 1,818
Brig. Gen. Henry H.
Lockwood
1st Md., Eastern Shore, 532
1st Md., Potomac Home Bde., 674
150th N.Y., 609
3rd Bde., 1,598
Brig. Gen. Thomas H. Ruger
Col. Silas Colgrove
27th Ind., 339
2nd Mass., 316
13th N.J., 347
107th N.Y., 319
3rd Wisc., 260

2nd Div., 3,964
Brig. Gen. John W. Geary
Provost Gd., *28th Pa., B*, 37
1st Bde., 1,798
Col. Charles Candy
5th Ohio, 302
7th Ohio, 282

29th Ohio, 308
66th Ohio, 303
28th Pa., A, C-K, 303
147th Pa., A-H, 298
2nd Bde., 700
Col. George A. Cobham, Jr.
Brig. Gen. Thomas L. Kane
Col. George A. Cobham, Jr
29th Pa., 357
109th Pa., 149
11th Pa., 191
3rd Bde., 1,424
Brig. Gen. George S. Greene
60th N.Y., 273
78th N.Y., 198
102nd N.Y., 230
137th N.Y., 423
149th N.Y., 297

XII Corps Art. Bde., **20 guns/391**
Lt. Edward D. Muhlenberg
M, 1st N.Y. Lt. Art. [four 10-pdr. Parrotts], 90
E, Pa. Lt. Art. [six 10-pdr. Parrotts], 139
F, 4th U.S. Art. [six 12-pdr. Napoleons], 89
K, 5th U.S. Art. [four 12-pdr. Napoleons], 77

Cavalry Corps, 52 guns/12,101

Maj. Gen. Alfred Pleasonton
Escort: *6th U.S. Cav.*, 471

1st Div., 4,544
Brig. Gen. John Buford
1st Bde., 1,600
Col. William Gamble
HQ Gd., *6th N.Y. Cav., L*, 35
8th Ill. Cav., 470
12th Ill. Cav., E-F, H-I, 233
3rd Ind. Cav., G-M, 313
8th N.Y. Cav., 580
2nd Bde., 1,148
Col. Thomas C. Devin
6th N.Y. Cav., B-C, E-G, I, M, 215

9th N.Y. Cav., 367
17th Pa. Cav., A-C, E-G, I-M, 464
3rd W. Va. Cav., A, C, 59
Res. Bde. [3 July with 3rd Cav Div.], **1,792**
Brig. Gen. Wesley Merritt
6th Pa. Cav., A-D, F-H, K-M, 242
1st U.S. Cav., A-E, G-M, 362
2nd U.S. Cav., 407
5th U.S. Cav., 306

2nd Div. 2,664
Brig. Gen. David. McM. Gregg
HQ Gd., *1st Ohio Cav., A*, 37

1st Bde., 2 guns/1,561
    Col. John B. McIntosh
    1st Md. Cav., A-L, 285
    Purnell (Md.) Legion, A, 66
    1st Mass., A-H [Guarding army HQ], 250
    1st Pa. Cav., A-G, I-M, 355
    3rd Pa. Cav., 335
    H, 3rd Pa. Heavy Art [two 3"-rifles], 52*
2nd Bde. #
    Col. Pennock Huey
    2nd N.Y. Cav.
    4th N.Y. Cav.
    6th Ohio Cav., A-E, G-L
    8th Pa. Cav.
3rd Bde., 1,263
    Col. J. Irvin Gregg
    1st Maine Cav., A-B, D-K, M, 315
    10th N.Y. Cav., 333
    4th Pa. Cav., 258
    16th Pa. Cav., 349

3rd Div., 3,902
    Brig. Gen. Judson Kilpatrick
    HQ Gd., *1st Ohio Cav., C*, 40
1st Bde., 1,925
    Brig. Gen. Elon Farnsworth (k)
    Col. Nathaniel P. Richmond
    5th N.Y. Cav., 420

    18th Pa. Cav., 509
    1st Vt. Cav., 600
    1st W. Va. Cav., B-H, L-N, 395
2nd Bde. [3 July with 2nd Cav. Div.], **1,934**
    Brig. Gen. George A. Custer
    1st Mich. Cav., 427
    5th Mich. Cav., 646
    6th Mich. Cav., 477
    7th Mich. Cav., A-K, 383

1st Horse Art. Bde., 28 guns/492
    Capt. James M. Robertson
    9th Mich. Btty [six 3"-rifles], 111
    6th N.Y. Btty [six 3"-rifles], 103
    B/L, 2nd U.S. Art. [six 3"-rifles], 99
    M, 2nd U.S. Art. [six 3"-rifles], 117
    E, 4th U.S. Art. [four 3"-rifles], 60

2nd Horse Art. Bde., 12 guns/272
    Capt. John C. Tidball
    E/G, 1st U.S. Art. [four 3"-rifles], 82
    K, 1st U.S. Art. [six 3"-rifles], 114
    A, 2nd U.S. Art. [six 3"-rifles], 74
    C, 3rd U.S. Art. [six 3"-rifles] #

Artillery Reserve, 110 guns/2,376

    Brig. Gen. Robert O. Tyler
    HQ Gd., *32nd Mass., C*, 45
    Train Gd., *4th N.J., B, D-G, I-K*, 273
    Ordnance Corps, 11
1st Reg. Bde., 24 guns/445
    Capt. Dunbar R. Ransom (w)
    H, 1st U.S. Art. [six 12-pdr. Napoleons], 129
    F/K, 3rd U.S. Art. [six 12-pdr. Napoleons]
    F/K, 3rd U.S. Art. [six 12-pdr. Napoleons], 115

    C, 4th U.S. Art. [six 12-pdr. Napoleons], 95
    C, 5th U.S. Art. [six 12-pdr. Napoleons], 104
1st Vol. Bde., 22 guns/385
    Lt. Col. Freeman McGilvery
    5th Btty, Mass. Lt. Art/10th N.Y. Btty [six 3"-rifles] 104
    9th Btty, Mass. Lt. Art. [six 12-pdr. Napoleons], 104
    15th Btty, N.Y. Lt. Art. [four 12-pdr. Napoleons], 70
    C/F, Pa. Lt. Art. [six 3"-rifles], 105

2nd Vol. Bde., 12 guns/241
Capt. Elijah D. Taft
B, 1st Conn. Heavy Art. [four 4.5-inch rifles] #
M, 1st Conn. Heavy Art. [four 4.5-inch rifles] #
2nd Btty, Conn. Lt. Art. [four 14-pdr. James' & two 12-pdr howitzers], 93
5th Btty, N.Y. Lt. Art. [six 20-pdr. Parrotts], 146

3rd Vol. Bde., 22 guns/431
Capt. James F. Huntington
1st Btty, N.H. Lt. Art. [six 3"-rifles], 86
H, 1st Ohio Lt. Art. [six 3"-rifles], 99

F/G, 1st Pa. Lt. Art. [six 3"-rifles], 144
C, W. Va. Lt. Art. [four 10-pdr. Parrotts], 100

4th Vol. Bde., 24 guns/499
Capt. Robert H. Fitzhugh
6th Btty, Maine Lt. Art. [six 12-pdr. Napoleons], 87
A, Md. Lt. Art. [six 3"-rifles], 106
1st Btty, N.J. Lt. Art. [six 10-pdr. Parrotts], 98
G, 1st N.Y. Lt. Art. [six 12-pdr. Napoleons], 84
K, 1st N.Y. Lt. Art./11th N.Y. Btty [six 3"-rifles], 128

Notes to the Order of Battle for the *Army of the Potomac*

1. The *Engineer Brigade* was at Beaver Dam Creek, Maryland, on 1 July when, with the exception of the *U.S. Engineer Battalion*, it was ordered to Washington, arriving there on 3 July.

2. The *12th* and *15th Vt.* were guarding trains during the battle.

3. The *102nd Pa.* was guarding trains at Westminster during the battle, but on 3 July a detachment of 103 men drawn from various companies arrived on the battlefield escorting a supply column and was immediately put into the lines.

4. The *84th Pa.*, Btty. *M, 1st Conn. Heavy Art.*, *2nd Vol. Art. Bde.*, the *2nd Bde.* of the *2nd Cav. Div.*, and Btty. *C, 3rd U.S. Art.* of the *3rd Cav. Div.* were guarding trains at Westminster during the battle.

5. *Btty. B, 1st Conn. Heavy Art.*, *2nd Vol. Arty. Bde.*, was at Taneytown during the battle.

The Army of Northern Virginia, 280 guns/69,915

Headquarters and Staff, 17
Commanding General: Gen.
Robert E. Lee
Chief-of-Staff & Inspector
General: Col. R. H. Chilton
Chief-of-Artillery: Brig. Gen.
William N. Pendleton
Chief-of-Engineers (Acting):
Maj. Gen. Isaac Trimble
(w)
Chief-of-Ordnance: Lt. Col.
Briscoe G. Baldwin
Chief-of Commissary: Lt. Col.
Robert G. Cole
Chief Quartermaster: Lt. Col.
James L. Corley

Judge Advocate General: Maj.
H.E. Young
Military Secretary (Acting
Assistant
Chief-of-Artillery): Col.
A.L. Long
Aide-de-camp and Assistant
Military Secretary: Maj.
Charles Marshall
Aide-de-camp and Assistant
Inspector General: Maj.
Charles S. Venable
Staff Engineer: Capt. S.R.
Johnston
Medical Director: Dr.
Lafayette Guild
Escort: 39th Va. Cav. Bn., C, 43

First Corps, 87 guns/20,811

Lt. Gen. James Longstreet

McLaws' Div., 16 guns/6,924
Maj. Gen. Lafayette McLaws
Kershaw's Bde., 2,183
Brig. Gen. Joseph B. Kershaw
2nd ["Palmetto"] S.C., 412
3rd S.C., 406
7th S.C., A-M, 408
8th S.C., A-M, 300
15th S.C., 448
3rd ["James'"] S.C. Bn., A-G, 203
Semmes' Bde., 1,334
Brig. Gen. Paul Jones Semmes
(mw)
Col. Goode Bryan
10th Ga., 303
50th Ga., 302
51st Ga., 303
53rd Ga., 422
Barksdale's Bde., 1,620
Brig. Gen. William Barksdale
(mw)
Col. B.G. Humphreys
13th Miss., 481
17th Miss., 469

18th Miss., 242
21st Miss., A & C-L, 424
Wofford's Bde., 1,398
William T. Wofford
16th Ga., 393
18th Ga., 302
24th Ga., 303
Cobb' (Ga.) Legion, A-G, 213
Phillip's (Ga.) Legion, A-F, L-M,
& O, 273
Div. Art. Bn., 16 guns/378
Col. H.C. Cabell
Pulaski (Ga.) Btty [two 3"-rifles &
two 10-pdr. Parrotts] 63
Troup County (Ga.) Lt. Btty.
[two 12-pdr. howitzers & two
10-pdr. Parrotts], 900
A ["Ellis Lt."], 1st N.C. Art. [two
12-pdr. Napoleons & two 3"-
rifles], 131
1st Richmond (Va.) Howitzer
Btty. [two 12-pdr. Napoleons
& two 3"-rifles], 90

Pickett's Div., 18 guns/5,578
Maj. Gen. George E. Pickett

Garnett's Bde., 1,459
Brig. Gen. Richard B. Garnett (k)
Maj. Charles S. Peyton
8th Va., 193
18th Va., 312
19th Va., 328
28th Va., A-G & I-K, 333
56th Va., 289

Kemper's Bde., 1,634
Brig. Gen. James L. Kemper (w, c)
Col. Joseph Mayo, Jr. (w)
1st Va. ["Williams' Rifles"], B-D & G-I, 209
3rd Va., 332
7th Va., A-G & I-K, 335
11th Va., 359
24th Va., 395

Armistead's Bde., 2,055
Brig. Gen. Lewis A. Armistead (k)
Col. W. R. Aylett (w)
9th Va., A-G & I-K, 318
14th Va., 422
38th Va., 400
53rd Va., 435
57th Va., 476

Div. Art. Bn., 18 guns/419
Maj. James Dearing
Fauquier (Va.) Btty. [Formerly G, 49th Va. Inf.] [four 12-pdr. Napoleons & two 10-pdr. Parrotts], 134
Lynchburg (Va.) Btty. [four 12-pdr. Napoleons], 96
Richmond "Fayette" (Va.) Btty. [two 12-pdr. Napoleons & two 10-pdr. Parrotts], 90
Richmond "Hampden" (Va.) Btty. [two 12-pdr. Napoleons, one 3"-rifle, & one 10-pdr. Parrott], 90

Hood's Div., 19 guns/7,375
Maj. Gen. John B. Hood (w)
Brig. Gen. Evander McI. Law

Law's Bde., Brig. Gen. Evander McI. Law, 1,933
Col. James L. Sheffield
4th Ala., 308
15th Ala., A-L, 499
44th Ala., 363
47th Ala., 347
48th Ala., 374

Robertson's Bde., 1,734
Brig. Gen. Jerome B. Robertson (w)
3rd Ark., 479
1st Tex., A-M, 426
4th Tex., 415
5th Tex., 409

Anderson's Bde., 1,874
Brig. Gen. George T. Anderson (w)
Lt. Col. William Luffman
7th Ga., 377
8th Ga., 312
9th Ga., B-K, 340
11th Ga., 310
59th Ga., 525

Benning's Bde., 1,420
Brig. Gen. Henry L. Benning
2nd Ga., 348
15th Ga., 368
17th Ga., 350
20th Ga., 350

Div. Art. Bn., 19 guns/403
Maj. M. W. Henry
D [Rowan], 1st N.C. Art. [two 12-pdr. Napoleons, two 10-pdr. Parrotts, two 3"-rifles], 148
F [Branch's], 13th N.C. Art. Bn. [one 6-pdr., one 12-pdr. howitzer, three 12-pdr. Napoleons], 112
Charleston (S.C.) "German" Lt. Btty [four 12-pdr. Napoleons], 71
Palmetto (S.C.) Lt. Btty. [two 12-pdr. Napoleons & two 10-pdr. Parrotts], 63

First Corps Res. Art., 34 guns/918

Col. J. B. Walton

Alexander's Art. Bn., 24 guns/576

Col. Edward P. Alexander

Madison (La.) Btty. [four 24-pdr. howitzers], 135

Brooks (S.C.) Lt. Btty. [four 12-pdr. howitzers], 71

Ashland (Va.) Btty. [two 12-pdr. Napoleons & two 10-pdr. Parrotts], 103

Bath (Va.) Btty. [four 12-pdr. Napoleons], 90

Bedford (Va.) Btty. [four 3"-rifles], 78

Richmond "Parker" (Va.) Btty. [three 3"-rifles & one 10-pdr. Parrott], 90

Washington Art. Bn. (La.), 10 guns/338

Maj. B.F. Eshleman

1st Coy. [one 12-pdr. Napoleon], 77

2nd Coy. [two 12-pdr. Napoleon & one 12-pdr. howitzer], 80

3rd Coy. [three 12-pdr. Napoleons], 92

4th Coy. [two 12-pdr. Napoleons & one 12-pdr. howitzers], 80

Second Corps, 78 guns/20,572

Lt. Gen. Richard S. Ewell

Escort, 39th Va. Cav. Bn., C, 31

Early's Div., 16 guns/5,460

Maj. Gen. Jubal A. Early

Hays' Bde., 1,295

Brig. Gen. Harry T. Hays

5th La., 196

6th La., 218

7th La., 235

8th La., 296

9th La., A-I, 347

Smith's Bde., 806

Brig. Gen. William Smith

31st Va., 267

49th Va., A-F & H-K, 281

52nd Va., 254

Hoke's Bde., 1,244

Col. Isaac E. Avery (mw)

Col. A.C. Godwin

6th N.C., 509

21st N.C., A, C-D, & F-M, 436

57th N.C., 297

Gordon's Bde., 1,813

Brig. Gen. John B. Gordon

13th Ga., 312

26th Ga., 315

31st Ga., 252

38th Ga., 341

60th Ga., 299

61st Ga., 288

Div. Art. Bn., 16 guns/290

Lt. Col. H. P. Jones

Louisiana Gd. Btty. [two 3"-rifles & two 10-pdr. Parrotts], 60

Charlottesville (Va.) Btty. [four 12-pdr. Napoleons], 71

Richmond "Courtney" (Va.) Btty. [four 3"-rifles], 90

Staunton (Va.) Btty. [four 12-pdr. Napoleons], 60

Johnson's Div., 16 guns/6,433

Maj. Gen. Edward Johnson

Steuart's Bde., 2,121

Brig. Gen. George H. Steuart

1st Md. Bn., A-G, 400

1st N.C., 377

3rd N.C., 548

10th Va., A-L, 276

23rd Va., 251

37th Va., A-F, G/I, & H-K, 264

Nicholls' ["Louisiana Tigers"] Bde., 1,104

Col. J.M. Williams

1st La., A-G & I-K, 172

2nd La., 236

10th La., 226

14th La., 281
15th La., 186

The "Stonewall" Bde., 1,323
  Brig. Gen. James A. Walker
  2nd Va., 333
  4th Va., A-I & L, 257
  5th Va., A & C-L, 345
  27th Va., B-H, 148
  33rd Va., 236

Jones' Bde., 1,520
  Brig. Gen. John M. Jones (w)
  Lt. Col. R.H. Dungan
  21st Va., A & C-K, 236
  25th Va., 280
  42nd Va., 265
  44th Va., B-K, 227
  48th Va., 265
  50th Va., 240

Div. Art. Bn., 16 guns/356
  Maj. J.W. Latimer (mw)
  Capt. C.I. Raine
  1st Md. Btty. [four 12-pdr. Napoleons], 90
  4th Md. ["Chesapeake"] Btty. [four 10-pdr. Parrotts], 76
  Allegheny (Va.) Btty. [two 12-pdr. Napoleons & two 3"-rifles], 91
  Lynchburg "Lee" (Va.) Btty. [one 3"-rifle, one 10-pdr. Parrott, & two 20-pdr. Parrotts], 90

Rodes' Div., 16 guns/7,986
  Maj. Gen. Robert E. Rodes

Daniel's Bde., 2,162
  Brig. Gen. Junius Daniel
  32nd N.C., A-B & D-K, 454
  43rd N.C., 572
  45th N.C., 570
  53rd N.C., A-B & D-K, 322
  2nd N.C. Bn., A-B & D-H, 240

Iverson's Bde., 1,384
  Brig. Gen. Alfred Iverson
  5th N.C., 473
  12th N.C., 219
  20th N.C., 372
  23rd N.C., 316

Doles' Bde., 1,323
  Brig. Gen. George Doles
  4th Ga., 341
  12th Ga., 327
  21st Ga., 287
  44th Ga., 364

Ramseur's Bde., 1,027
  Brig. Gen. Stephen Ramseur
  2nd N.C., 243
  4th N.C., 196
  14th N.C., 306
  30th N.C., 278

O'Neal's Bde., 1,688
  Col. Edward A. O'Neal
  3rd Ala., A-L, 350
  5th Ala., 317
  6th Ala., 382
  12th Ala., 317
  26th Ala., 319

Div. Art. Bn., 16 guns/385
  Lt. Col. Thomas H. Carter
  Jefferson Davis (Ala.) Art. [four 3"-rifles], 79
  King William (Va.) Btty. [two 12-pdr. Napoleons & two 10-pdr. Parrotts], 103
  Morris (Va.) Btty [four 12-pdr. Napoleons], 114
  Richmond "Orange" (Va.) Btty. [two 3"-rifles & two 10-pdr. Parrotts], 80

Second Corps Res. Art., 30 guns/648
  Col. J. Thompson Brown

1st Va. Art. Bn., 20 guns and 367
  Capt. Willis J. Dance
  Powhatan (Va.) Btty. [four 3"-rifles], 78
  2nd Richmond (Va.) Howitzer Btty. [four 10-pdr. Parrotts], 64
  3rd Richmond (Va.) Howitzer Btty. [four 3"-rifles], 62
  1st Rockbridge (Va.) Btty. [four 20-pdr. Parrotts], 85

Salem (Va.) "Flying" Btty. [two 12-pdr. Napoleons & two 3"-rifles], 66

Nelson's Art. Bn., 10 guns/277
Lt. Col. William Nelson
Ga. Reg. Btty [two 3"-rifles & one 10-pdr. Parrott], 73

Amherst (Va.) Btty. [three 12-pdr. Napoleons], 105
Fluvanna (Va.) "Consolidated" Btty [three 12-pdr. Napoleons & one 3"-rifle], 90

Third Corps, 84 guns/22,083

Lt. Gen. Ambrose P. Hill

Heth's Div., 15 guns/7,461
Maj. Gen. Henry Heth (w)
Brig. Gen. James J. Pettigrew (w)

1st Bde., 2,584
Brig. Gen. James J. Pettigrew (w)
Col. J.K. Marshall (w, c)
11th N.C., 617
26th N.C., 843
47th N.C., 567
52nd N.C., 553

2nd Bde., 971
Col. J.M. Brockenbrough
40th Va., 253
47th Va., A-G & H-I, 209
55th Va., A & C-M, 268
22nd Va. Bn., A-B D-E, & G-H, 237

3rd Bde., 1,197
Brig. Gen. James J. Archer (c)
Col. B.D. Fry (w, c)
Lt. Col. S.G. Shepard
13th Ala., 308
5th Ala. Bn., 135
1st [Turney's] Tenn. (Provisional Army), 281
7th Tenn., 249
14th Tenn., A-E & G-L, 220

4th Bde., 2,305
Brig. Gen. Joseph R. Davis
2nd Miss., A-L, 492
11th Miss., 592
42nd Miss., 575
55th N.C., 640

Div. Art. Bn., 15 guns/396
Lt. Col. John J. Garnett
Donaldsville (La.) Btty. [two 3"-rifles & one 10-pdr. Parrott], 114
Norfolk "Huger's" (Va.) Btty. [two 12-pdr. Napoleons, one 3"-rifle, & one 10-pdr. Parrott], 77
Norfolk "Lt. Art. Blues" (Va.) Btty. [H, 16th Va] [two 3"-rifles & two 12-pdr. howitzers], 106
Pittsylvania (Va.) Btty. [two 12-pdr. Napoleons & two 3"-rifles] 90

Pender's Div., 16 guns/6,735
Maj. Gen. William D. Pender (mw)
Brig. Gen. James H. Lane
Maj. Gen. Isaac Trimble (w, c)
Brig. Gen. James H. Lane

1st Bde.
Col. Abner Perrin, 1,882
1st S.C. (Provisional Army), A-C & E-L, 328
1st S.C. Rifles, A-H & K-L, 366
12th S.C., 366
13th S.C., 390
14th S.C., 428

2nd Bde., 1,734
Brig. Gen. James H. Lane
Col. C.M. Avery
Brig. Gen. James H. Lane (w)
Col. C.M. Avery
7th N.C., 291
18th N.C., 346

28th N.C., 346
33rd N.C., 368
37th N.C., 379
3rd Bde., 1,326
Brig. Gen. Edward Thomas
14th Ga., 331
35th Ga., 331
45th Ga., 331
49th Ga., 329
4th Bde., 1,405
Brig. Gen. Alfred M. Scales (w)
Lt. Col. G.T. Gordon
Col. W. Lee J. Lowrance (w)
13th N.C., 232
16th N.C., B-K & M, 321
22nd N.C., A-M, 321
34th N.C., 311
38th N.C., 216
Div. Art. Bn., 16 guns/377
Maj. William T. Poague
C, 1st N.C. Art. ["Charlotte Btty."] [two 12-pdr. Napoleons & two 12-pdr. howitzers], 125
Madison (Miss.) Btty. [three 12-pdr. Napoleons & one 12-pdr. howitzer], 91
Albemarle "Everett" (Va.) Btty [two 3"-rifles, one 10-pdr. Parrott, & one 12-pdr. howitzer], 94
Warrenton (Va.) Btty [two 12-pdr. Napoleons & two 12-pdr. howitzers], 58

Anderson's Div., 17 guns/7,135
Maj. Gen. Richard H. Anderson
Wilcox's Bde., 1,726
Brig.Gen. Cadmus M. Wilcox
8th Ala., 477
9th Ala., 306
10th Ala., 311
11th Ala., 311
14th Ala., 316

Mahone's Bde., 1,542
Brig. Gen. William Mahone
6th Va., 288
12th Va., 348
16th Va., A-G, 270
41st Va., 276
61st Va., 356
Perry's ["Florida"] Bde.
Col. David Lang, 742
2nd Fla., A-M, 242
5th Fla., 321
8th Fla., 176
Wright's Bde., 1,413
Brig. Gen. Ambrose R. Wright
Col. William Gibson (w, c)
Brig. Gen. Ambrose R. Wright
3rd Ga., A-L, 441
22nd Ga., 400
48th Ga., 395
2nd Ga. Bn, 173
Posey's Bde., 1,322
Brig. Gen. Carnot Posey
12th Miss., 305
16th Miss., 385
19th Miss., 372
48th Miss., 256
Div. Art. Bn. [11th Ga. Arty Bn., "The Sumpter Artillery"], 17 guns/384
Maj. John Lane
A Coy. [one 12-pdr. howitzer, one 12-pdr. Napoleon, one 3"-rifle and one 10-pdr. Parrott], 130
B Coy. [four 12-pdr. howitzers & two 12-pdr. Napoleons], 124
C Coy. [three 3"-Navy rifles & two 10-pdr. Parrotts], 121

Third Corps Res. Art., 36 guns/736
Col. R. Lindsay Walker
McIntosh's Art. Bn., 16 guns/357
Maj. D.G. McIntosh

Hardaway (Ala.) Btty. [two 3" rifles & two 12-pdr. Whitworths], 71

Danville (Va.) Btty. [four 12-pdr. Napoleons], 114

2nd Rockbridge (Va.) Btty. [two 12-pdr. Napoleons & two 3"-rifles], 67

Johnson's Richmond (Va.) Btty. [four 3"-rifles], 96

Pegram's Art. Bn., 20 guns/375
Maj. W.J. Pegram

Pee Dee (S.C.) Btty. [D, 1st S.C.] [four 3"-rifles], 65

Fredericksburg (Va.) Btty. [two 12-pdr. Napoleons & two 3"-rifles], 71

Richmond "Crenshaw" (Va.) Btty. [two 12-pdr. Napoleon's and two 12-pdr. howitzers], 76

Richmond "Letcher" (Va.) Btty. [two 12-pdr. Napoleons & two 10-pdr. Parrotts], 65

Richmond "Purcell" (Va.) Btty. [four 12-pdr. Napoleons], 89

Cavalry Div. 17 guns/6,389
Maj. Gen. James E. B. Stuart

Hampton's Bde., Brig. Gen. Wade Hampton (w), 1,751
Col. Lawrence S. Baker

Cobb's (Ga.) Cav. Legion, A-L, 330

Phillip's (Ga.) Cav. Legion, A-G, 238

Jeff Davis (Miss.) Cav. Legion, A-F, 246

1st N.C. Cav., A-K, 407

1st S.C. Cav., A-K, 339

2nd S.C. Cav., A-K, 186

F. Lee's Bde., 1,913
Brig. Gen. Fitzhugh Lee

1st Md. Cav. Bn., A-E [with II Corps], 310

1st Va. Cav., A-K, 310

2nd Va. Cav., A-K, 385

3rd Va. Cav., A-K, 210

4th Va. Cav., A-H & K, 544

5th Va. Cav., A-K, 150

W. H. F. Lee's Bde., 1,173
Col. John R. Chambliss, Jr.

2nd N.C. Cav., A-K, 145

9th Va. Cav., A-K, 490

10th Va. Cav., A-K, 236

13th Va. Cav., A-K, 298

Jenkins' Bde.,2 guns/1,126
Brig. Gen. Albert G. Jenkins (w),
Col. M.J. Ferguson

14th Va. Cav., A-L, 265

16th Va. Cav., A-L, 265

17th Va. Cav., A-K, 241

34th Va. Cav. Bn., A-G, 172

36th Va. Cav. Bn., A-E, 125

Charlottesville (Va.) Horse Art. Bn. [two 12 pdr. howitzers], 54

Robertson's Bde. #
Brig. Gen. Beverly H. Robertson

4th N.C. Cav., A-H

5th N.C. Cav., A-K

Jones' Bde. #
Brig. Gen. William E. Jones

6th Va. Cav., A-K

7th Va. Cav., A-K

11th Va. Cav., A-K

35th Va. Cav. Bn.

Div. Horse Art. Bn., 15 guns/406
Maj. R.F. Beckham

2nd Baltimore (Md.) Lt. Art. Btty. [four 10-pdr. Parrotts], 106

Ashby (Va.) Horse Art. Btty. [four pieces, type unknown] #

1st Stuart (Va.) Horse Art. Btty. [four 3"-rifles], 106

Lynchburg (Va.) Horse Art.
Btty. [four pieces, type un-
known] #
2nd Stuart (Va.) Horse Art.
Btty. [two 3"-rifles & two 12-
pdr. Napoleons], 106
Washington (S.C.) Btty. [two 3"-
rifles & two 12-pdr. Napole-
ons], 79

Imboden's Command #
Brig. Gen. John D. Imboden
18th Va. Cav., A-K
62nd Va. Mounted Inf., A-M
Va. Partisan Rangers, McNeil's
Coy.
Staunton (Va.) Horse Art. Btty
[six 12-pdr. Napoleons?]

Notes to the Order of battle of the Army of Northern Virginia

1. The cavalry brigades of Brig. Gen. Beverly H. Robertson and Brig. Gen. William E. Jones spent most of the campaign on line of communications and security duty. The 35th Va. Cav. Bn. of Jones' brigade remained in Virginia during the campaign.

2. The Ashby (Va.) and the Lynchburg (Va.) horse artillery batteries were with Robertson's and Jones' brigades during the campaign.

3. The command of Brig. Gen. John D. Imboden was not technically a part of the Army of Northern Virginia, but had been ordered to cooperate with it. Not engaged at Gettysburg, the brigade spent the campaign covering the rear of the army against possible threats from West Virginia and Western Pennsylvania.

Index